Space and Food in the City

Alec Thornton

Space and Food in the City

Cultivating Social Justice and Urban Governance
through Urban Agriculture

Alec Thornton
The University of New South Wales
UNSW Canberra, ACT, Australia

ISBN 978-3-319-89323-5   ISBN 978-3-319-89324-2   (eBook)
https://doi.org/10.1007/978-3-319-89324-2

Library of Congress Control Number: 2018938333

Cover illustration: © nemesis2207/Fotolia.co.uk

Printed on acid-free paper

This Palgrave Pivot imprint is published by the registered company Springer International Publishing AG part of Springer Nature
The registered company address is: Gewerbestrasse 11, 6330 Cham, Switzerland

CONTENTS

LIST OF TABLES

SUMMARY

Urban social movements concerned with social equity, environmental and food justice themes are emerging as influential agents in shaping good urban governance outcomes. These outcomes include improved interconnectivity among city–community activists, public and private stakeholders in local urban food systems and re-prioritising urban planning to accommodate and promote activities that improve urban resilience, social cohesion and local food economies. Alongside rapid urban population growth, various forms of urban—and peri-urban—agriculture (UA) activity, such as community and market gardens and related urban food networks, such as farmer's markets and community-supported food-box schemes are also expanding in cities, globally. Although UA routinely struggles to penetrate public policy space, it has made notable contributions to urban food supply, particularly during times of crisis—in North and South cities. Increasingly, UA is contributing to urban policy spaces, largely in western cities, to address the challenges of urbanisation and climate change.

This book will engage with two key themes, first, grassroots and municipal-level interest in the potential of UA to address socio-spatial inequalities are explored, and will feature a case study of urban activism and community gardens in Berlin, Germany. Second, the book will attempt to highlight similarities and distinctions between UA in developed (North) and developing (South) countries, with a focus on sub-Saharan Africa—a region that lagged behind other developing economies of the world in meeting the Millennium Development Goals

(2000–2015), especially in terms of poverty, food security and job creation—an outcome that is likely to repeat with the more extensive Sustainable Development Goals. This book will emphasise the importance of a culture of mobilised citizens, as knowledgeable urban activists, to demand policy action for urban sustainability. Ultimately, such progressive urban activism for social and spatial equity is limited in cities in the South. A culture of urban activism is critical, as they form the critical mass of grassroots agents required for influencing change in urban and regional policy for resilient urban food systems. In exploring these themes, this book will draw from case study examples from my own research, as well as others.

CHAPTER 1

Introduction

Abstract This chapter will provide an overview of the main theoretical frameworks to be explored in this book, which are urban governance and social mobility, particularly in the production of urban food spaces. These frameworks will provide the analytical lens to explore city–community engagement (or lack thereof) in urban agriculture (UA) and urban food security in cities in developed and developing countries. This chapter will discuss ideas in critical urban theory, with respect to social production of local food space, food security, with a focus on current research to be explored in the literature on UA from cities in North and South contexts.

Keywords Urban governance · Social mobility · Food security
Social movements · Global North · Global South

INTRODUCTION

Overview

Urban social movements concerned with social equity, environmental and food justice themes are emerging as influential agents in shaping good urban governance outcomes. These outcomes include improved interconnectivity among city–community activists, public and private stakeholders in local urban food systems and re-prioritising urban planning to accommodate and promote activities that improve urban resilience, social cohesion and local food economies. Alongside rapid urban population growth, various forms of urban agriculture (UA) activity, such as community and market gardens and farmers' markets are also expanding, in the so-called 'Global North and South'.

For nearly a decade, I've explored UA conceptually, as well as its various manifestations, scale, practice and utility in urban areas in developed and developing countries. During this time, I have recognised growing sociopolitical interest in the potential of UA to strengthen social cohesion and improve urban health and food knowledge (the latter is an often-neglected aspect of food security) in marginalised communities. Secondly, although UA exists in developed and developing countries, it is worthwhile to recognise its distinctions, in theory and practice, in 'North and South' cities. These distinctions are important, as issues of social equity, food justice and inclusive approaches to sustainable urban policymaking play out differently (if at all, in some contexts) in urban policy spaces and in parallel with an urbanising global population.

Particular benefits of the content of this book are largely offered through extensive case study examples, which are drawn from my own research, as well as others, on experiences of UA social equity and urban governance in cities in developing and developed countries. Admittedly, while attempting to offer some form of critique of these experiences from Global North and South perspectives is daunting, I believe it is still worthwhile. In guiding this analysis, I adopt urban governance and social mobility frameworks. This joins together structural, cultural and rational actor approaches to cross-comparison. Discussed below, UN-Habitat offers a good overview of urban governance, which is a broad concept that includes the role of institutions and individuals in creating an enabling environment. While I certainly agree that poverty and food survival challenges differ in scope and scale in a North and South comparison (and absolute vs relative poverty), the bigger picture of expanding city populations and related challenges to food security impacts both contexts. From urban governance discourses, how North and South cities respond to or facilitate urban-socio mobility, as people seek to meet various needs, such as fresh food access and availability, will feature in the case study discussion in a later chapter. Responses will differ, as agendas and participants' do differ—locally and globally—leading to different UA outcomes.

In this introductory chapter, ideas in critical urban theory, with respect to social production of local food space, food security, with a focus on current research will be explored in the literature on UA from cities in North and South contexts. Mentioned above, these experiences will be analysed using urban governance and social movement frameworks. In Chapter 2, following an overview of UA, distinctions and similarities in UA concepts and praxis will be identified between North and South contexts, where social equity (fairness), in terms of food justice, and inclusivity in urban socio-economic policy are highly significant. In Chapter 3, case study examples are drawn from my own research, as well as others, to discern to what extent UA in cities in developing and developed countries are creating spaces of social and food justice, and related challenges and opportunities for cities to provide and communities to access space. Chapter 4 will conclude with a discussion of socially produced spaces and food justice through UA. What are the possibilities and constraints of urban food movements in claiming their 'rights to the city' in the North and South?

Insights from my research experiences will highlight issues related to 'good' urban governance and urban social movements leading to city–community partnerships in planning for local and regional food systems. Normative or conventional voices claim that UA has limited economic value and increasing urban densities are inherently bad news for UA. These voices are increasingly becoming crowded out by success stories of community-driven urban food policy change. Its acceptance as a normative challenge is more political or ideological, than a real or physical obstacle to change in the urban food system. Through the experiences of social movements in producing urban food spaces under the auspices of 'good' and 'bad' urban governance, we can critique the potential for communities and local city councils to collaborate for reimagining urban spaces that emphasise social, economic and environmental justice for sustainable urban development, as opposed to 'actually existing neoliberalism' (Brenner and Theodore 2002; Brenner 2012; Okereke 2007; Eizenberg 2012).

Ultimately, this book is concerned with understanding the theoretical and applied implications of city–community interactions in UA and the processes by which alternative social-spatial production influences mainstream or broader perceptions and attitudes towards sustainable urban development.

When It Comes to Food Security, Is Urban Agriculture a 'Big Fail'?

UA, including peri-urban, is a broad research field that, since the 1980s, has grown into a field of multidisciplinary study. Earlier applied and thematic foci of UA had generated a wealth of knowledge on determining its socioeconomic nature and geographic extent, its impact on household food security and as an income source, its role in the informal economy and offering descriptions of various types of urban and peri-urban production systems (Thaman 1975; Sanyal 1985, 1987; Rogerson 1992; Smith and Tevera 1997; Mougeot 2000; Thornton 2008; Mun Bbun and Thornton 2013; Malan 2015). There is some debate on the actual impact of UA on food security, though these critiques tend to focus on food access and availability, while missing the significance of UA practices in improving and preserving food utilisation and knowledge. Following the 1996 World Food Summit in Rome, UN FAO defined food security as a situation that exists:

> [W]hen all people at all times have physical, social and economic access to sufficient, safe and nutritious food that meets their dietary needs and food preferences for an active and healthy life. (FAO 2015)

The FAO (2013) further explains food security as an outcome of the following four main dimensions:

- Access to food, or to what extent is access prohibited by political, social, cultural and economic factors
- Availability of food, or supply-side factors shaping sufficient quantities of food and at prices that people can afford
- Utilization of food, which considers dietary, nutritional knowledge and know-how of food preparation, as well as basic gardening skills to grow your own food (and, I would add, to teach these skills to others).
- Stability of the other three dimensions over time.

It is the 'use or knowledge of food' that is of particular interest here. For households experiencing high rates of urban food insecurity in South African cities (Johannesburg as high as 77%), food needs are largely met through the assistance of South Africa's substantial social welfare system to purchase food on the informal market (Thornton 2008, 2012; Battersby 2011, 2012; Malan 2015). While collecting household surveys during my fieldwork in South African townships, access to the cash economy through social welfare grants and the negative stigma attached by the youth to food growing (not 'modern' or 'something my grandparents had to do to survive', in Thornton 2012) would have a cumulative effect of diminishing the food growing knowledge base in urban areas.

Although frequently mentioned in UA research are the 'war gardens' (United States) of the First World War (WWI) and 'victory gardens' during Second World War (WWII) (United States and UK 'dig for victory') still offer important lessons in urban resiliency in the face of global crises (Miller 2003; Hayden-Smith 2014). These global conflicts brought immediate shortages of food, and other basic necessities. By the end of WWI, over 5 million gardens in the United States had produced $1.2 billion in foodstuffs. At the height of WWII (1943), there were 18 million gardens, with 12 million of these located in cities, producing 1/3 of all vegetables. By 1942, victory gardens (in cities and on farms) were producing 7.5 billion pounds of food (Lawson 2014: 182). Similarly, in the UK, nationwide campaigns for urban food growing produced up to 2 million tons of food during WWI and 1.3 million tons during WWII, which can be attributed to collective memory of food growing among the urban dwellers and the preservation of urban space to grow it (Barthel et al. 2015). Moreover, it was a time when "familiar self-sufficiency" served as a "powerful metaphor for freedom", and this was encouraged by the state (Miller 2003: 395).

Contemporary urban challenges brought by climate change and urban population expansion are placing the spectre of urban food insecurity on a 'slow boil', in contrast to global conflict impacts on supply lines. As inconceivable as a food supply crisis may appear, local governments could examine to what extent their cities and towns are food secure. The urban population has increased dramatically since the early to mid-twentieth century, when food growing knowledge was part of the collective memory and farms were more common closer to urban areas (thus beneficial for victory garden campaigns). In the last decade or so, skyrocketing prices of food resulting from the global food crisis (2007–2008), have raised public awareness of vulnerabilities in the global food system from climate change and biodiversity loss. For these reasons, and others no doubt, issues of food security, as well as the terms 'food deserts' and 'food sovereignty' have entered the mainstream public consciousness.[1] These issues are not unique to populations in developing countries, as 'food deserts' exist in western cities, typically found in low-income urban areas that have a higher density of fast-food restaurants and lack supermarkets or opportunities to locally (or conveniently) source fresh, nutritious produce (Walker et al. 2010). Currently, the loss of urban space to grow food would negatively affect the urban collective memory of growing food. What would the implications of this loss have on urban food security, particularly in an era of human-induced global warming and related volatility in the globalised food system? Is shortening the food supply chain, through UA, simply prudent food policymaking?

Social Movements and the Production of Space

Recently, UA is being discussed with a more critical lens (Eizenberg 2012; McClintock 2014; Thornton et al. 2018) and is increasingly viewed as an important component of local food systems for food sovereignty (Wittman et al. 2010), increasing urban densities and zoning considerations for urban food production and water resources to support it (Kühn 2003; Hodgson et al. 2011). Although there is no clear consensus on what defines a 'local' or 'regional' food system, it is generally viewed as geographically localised, as opposed to national or internally sourced. In describing the 'place of food', the constructed meaning of 'place', 'local' and 'community' are also spatial delimitations, which can lead to exclusionary orientations (Feagan 2007). In keeping these more negative tendencies in check, 'place' has a role in building alternative food systems, while also appealing to reflexive localism (DuPuis and Goodman 2005; Feagan 2007). As the global population increasingly urbanises, local urban food systems can reflect the sociocultural diversity of cities and re-localise what has become a 'placeless' globalised food system (O'Hara and Stagl 2001; Feagan 2007; Guthman 2008).

As a field of study, UA requires a more robust theoretical grounding to pull varied and localised insights into a clear ontological space for its critique, as an alternative and transformative social experience. This social

---

[1]Food sovereignty—the right of people to control their own food system—has been taken up by urban food activists in developing countries, as a concept that offers more potential than 'food security' to ensure access to nutritious and affordable local food (Wittman et al. 2010).

experience can be understood, as stated earlier by Lefebvre (1991), as the 'everyday life' of marginalised communities or groups seeking to reclaim (or appropriate) neoliberal (or dominant, market-oriented) urban spaces to meet a goal or purpose that is shared in common, such as equal access to affordable and nutritious food. This 'experience' would differ in the Global North and South contexts, which this book intends to provide some useful insights. For example, the fact that not all cities in the developing 'South' have experienced urban industrial-led growth, some argue that spatial inequality and related food insecurity exists as a product of postcolonialism (Sidaway 2000; Bek et al. 2004; Nally 2011). On the other hand, although South Africa is a highly industrialised country, city life does not evoke images of community-based urban activism, as the segregation policies of apartheid denied (at times forcefully) the possibility of such an urban culture from taking root. Since 1994, apartheid-era legacies contribute to a lack of community building and, more broadly, social exclusion, which is reinforced by neoliberal urban policy foci in South Africa's major cities (Beall 2002; Beall et al. 2014).

Although cities in the western world also experienced post-industrial decline (McCarthy 1997), the global financial and food (2007–2008) crisis cast a bright light on food systems failure and 'food deserts' in the North and South (Ghosh 2010; Rosin et al. 2013; Ledoux and Vojnovic 2013). In all cases, UA has emerged as an important strategy for some households for improving food security, as an income source and as a 'lived' social space produced by residents who share, in common, a desire to create spaces for social, economic and political equality. This view reflects Lefebvre's (1996 [1968], 1991) critique of space as socially and politically produced, which is partly constructed from his interpretation of Marx's historical materialism and dialectics, where spatial inequality is driven by scarcity, contradictions in the modes of production and a result of socially planned spaces where too much is allocated for the rich, while leaving too little for the poor (Elden 2004).

Although understandings of social movements and those that are of a particularly 'urban' nature do vary, this book adopts a broad view of a social movement as the mechanism through which actors engage in a collective action (Della Porta and Diani 2006). In the social movement literature, it is further described as a continuous interaction between challengers and power holders (Tilly 1999: 257). This interaction includes sustained public displays of a unified challenge or claim. The 'challenge' or 'claim', in this book, is people claiming their collective right to (neoliberal) urban space to develop or strengthen the local urban food system. This reflects Lefebvre's (2003 [1970]) 'right to the city' (discussed below) and Castells' (1983) ideas on 'urban' social movements, where an active civil society challenges dominant city planning regimes for the 'collective consumption' of resources, such as urban open green and vacant 'brownfield' space (Barthel et al. 2015). In social movement theory, cities have a role to play in the formation of social movements, as they are places of complex social and cultural relations that offer opportunities for strong coalitions among heterogeneous groups and organisations (Tilly 1999: 262). These coalitions are necessary to join-up 'strong' and 'weak' groups in sharing specialised knowledge or resources available to them (Nicholls 2008). These resources may then mobilise to campaign for shared goals or claim-making action to restructure urban space—a 'right to the city' (Brenner 2012). Dense urban social movements and urban governance are important elements of participatory urban democracy, where city–community partnerships can be drivers of urban change. Exploring the extent that these conceptual and applied urban processes are taking place (or not), in cities in the Global North and South, is a key objective of this book.

Urban Governance in the North and South

Urban governance is a broad concept that includes the role of institutions and individuals in creating an enabling environment, such as community–city partnerships, to effectively respond to the needs of all urban residents (Lindell 2008; UN-Habitat 2015). An enabling environment can include removing barriers to citizen mobility, as they engage in creative and innovative solutions in meeting a variety of needs (Healey 2004; Resnick 2014). This enabling environment reflects a shift, since the 1990s, to notions of urban 'governance' (as opposed to 'government'), which is concerned with the interplay between state (city/municipal) actors and society and the potential for collective city-citizen projects achieved through mobilising public and private resources (Pierre 2011: 5). This potential for city-citizen collective action is dependent on to what extent power relations can be negotiated, where voices and ideas from community-level activists are not only heard but contribute to decisions regarding urban socio-spatial policy and driving action for change (Eckhardt and Elander 2009: 14). For cities in the South, or developing countries, it is in this arena of urban governance where UA comes up short (Warshawsky 2014). Community-driven responses to food security, health and well-being struggle to emerge in Global South cities. Explored in Chapter 3, Global North cities typically fare much better in this space, as urban dwellers in western cities are typically more integrated in economic and social life, knowledgeable in policy space and more politicised thus engaged in local action in global problems (e.g. inequities connected to the globalised food system, negative impacts of climate change).

With supportive urban governance, perhaps UA can be one driver for enabling community–city partnerships for inclusive, resilient and liveable cities. Whereas concerns such as poverty, food access and availability and land pressures differ in scope and scale in a 'North-South' comparison (e.g. absolute, relative, food and time poverty), the bigger picture of expanding city population and related challenges to food supply/security affects both contexts. From urban governance discourses, this book will explore how Global North and South cities respond or facilitate urban-socio mobility, as people seek to meet various needs, such as fresh food access and availability. Responses will differ, as agendas and participants' can and do differ—locally and globally—leading to different and insightful UA outcomes.

Urban Agriculture and 'Rights to the City'

UA is increasingly viewed through the lens of critical urban theory (Eizenberg 2012; Crane et al. 2013), using ideas on production of social space, introduced by Lefebvre in the 1960s and advanced by contemporary scholars in critical urban geographies (Brenner and Elden 2001; Purcell 2002; Mitchell 2003; Harvey 2008; Marcuse 2009; Soja 2010; Brenner 2012). These ideas argue the importance of spatiotemporal relationships in the organisation and use of urban space, meaning how space is historically and socially configured and socially produced. In the 1960s, Lefebvre (1996 [1968]) argued that urban spaces in western civilisation are becoming increasingly commodified (reduction in public 'commons' for private consumption) and, in response, communities are claiming their 'rights to the city' thus giving rise to sociopolitical tension. This tension is driving concerns and political action for social and environmental justice (Harvey 2003, 2008). UA has emerged as a form of urban resistance among social movements in western cities that are demanding an equitable and accessible local food supply, clean environment through 'urban greening' and pedestrian-friendly and recreational inner-city zones. In western cities, UA has emerged as an important social movement that appears to extend beyond (though still inclusive of) food security, as it seeks to reconnect and enhance the symbiotic relationships between human and natural ecosystems.

UA, as an alternative and transformative system, has the potential to expand beyond the grassroots level to meet the needs of urban dwellers and the ecosystem (Roseland 2012; Thornton et al. 2012). Some western cities are identifying ways to integrate UA in urban planning (Kühn 2003; Land Stewardship Project-LSP 2010; American Planning Association [APA] 2010). The integration of environmentalism and neoliberalism is critiqued as a form of 'ecological modernisation' that may not lead to social change to more sustainable production and consumption, as theorised by Hajer (1995). Many critics argue that EM is little more than political 'green-washing', which does not address environmental degradation or the structural inequalities embedded produced by neoliberalism (Beck 1999; Foster 2002). Moreover, ecological modernisation is criticised as holding little relevance to developing countries, due to its emphasis on biotechnology and other innovative ecological solutions for sustainable western cities (Fisher and Freudenburg 2001).

Alternative spaces, by definition, are socially or community protected. It is not clear if the structural needs of the city and the social use-value (non-commodified) of alternative spaces can unite as complementary spaces, or will they compete for space (Harvey 2008, 2012). Some have argued that the contradictions between neoliberalism and alternative food systems, such as UA, need to be understood as both existing within the capitalist market logic and as a public good (McClintock 2014). Of secondary interest in this book are the processes whereby UA enters into formal policy planning, and to what extent are the possible preconditions for driving this transition likely to be highly localised and dependent on a citywide 'buy-in' of alternative socially defined approaches to fresh food access and availability, urban health and community building. Grassroots ideas and action on transforming the city for more equal social spaces appear to be converging in western cities. Discussed in a later Chapter (3), these ideas and action are influencing local government thinking and policy making for improving human–environment relationships in postindustrial cities.

The formation of urban social movements as influential political actors in policy change is dependent on the formation of dense linkages. This is likely the most significant hurdle for marginalised communities in cities in the Global South. This is not to say that social movements are completely non-existent in cities in developing countries. People do mobilise in protest of food system injustices. In 1977, mismanagement of the private and government food supply chain triggered the 'Egyptian Bread Riots', as people could not afford the price increases (of food that was available) that had resulted in cuts in food subsidies. Again, in 2007–2008, a global food crisis in the wake of an 83% increase in food price—caused by decline in grain stocks, oil price increase, speculation and adverse weather—were met with riots in cities across the globe (Mittal 2009). The disconnection of people from life-sustaining food, in the form of a global food supply and distribution system, places everyone at risk—citizens are powerless when disruptions occur (either natural or human-induced) to lengthy food supply chains (Bush 2010). People mobilising in response to government failure to ensure access and availability of food, is a form of a political social movement that emerges as a response or reaction to some form of injustice. However, as sustained organised agents for policy change, in the context of localising an urban food system, movements that are more 'reactionary' tend to lack the critical linkages needed to form a diverse coalition of actors (community, public and private) having various resources, knowledge and capabilities (organisational, vocational skills, policymaking, mobilising) to communicate across sectors and affect change.

Urban Agriculture and Systems Thinking

Contemporary approaches to systems thinking have evolved from its mechanistic origins (systems engineering, cybernetics, IT), to include studies of social, ecological, biological 'systems'. Post-war founders of The Society for General Systems Research (1954), Ludwig Von Bertalanffy, Kenneth Boulding, and others, recognised the potential for 'general systems theory' applications in a more comprehensive worldview, as an alternative to a reductionist methodology of isolated parts, as opposed to how these parts relate to the functioning whole (Von Bertalanffy and Rapoport 1963; Hammond 2003). A fundamental benefit of a systems thinking approach is to understand the interconnectivity of a collection of elements or parts and their effect on that 'system'. It is useful in helping various actors (or stakeholders) understand complex problems by linking it to 'the big picture'. Increasingly, systems thinking is being applied to cities, from urban ecosystems to urban social and ecological resilience (Adger 2000; Folke 2006; Tidball and Krasny 2014; Thornton 2008; Colding and Barthel 2013). In an era of global warming, resource scarcity and economic volatility, these 'big picture' issues seem to be baring down on cities—where the majority of an expanding global population live. What does systems thinking have to offer in working through problems associated with urbanisation and resource scarcity, particularly those resources tied to the global food system? Globally, urban dwellers are concerned about increasing cost of food and its relationship to the environmental cost of modern industrial agriculture and global food distribution systems.

In seeking policy change for urban food systems, where does UA fit in? How do various types of UA activities affect the wider urban food system? There are different classifications of agriculture, such as 'conservation', 'organic', 'commercial' and 'subsistence' agriculture. Urban horticulture also receives attention (Säumel et al. 2012), as its Latin origins *hortus* (garden) *cultura* (cultivation) suggest gardening in smaller spaces. Whereas *agri* (field) *cultura* (cultivation) denotes a jump in scale, thus implying broadacre 'field cultivation' of crops. Whereas horticulture is a branch of agriculture, contemporary use of the terms reveals considerable crossover between the two when discussing beneficial ornamental plants in urban areas to urban food production (Food and Agricultural Organisation [FAO] 2012). Broadly defined, 'agriculture' is the science or practice of farming, including cultivation of the soil for the growing of crops and the rearing of animals to provide food, wool and other products. In places where UA seems to flourish, it has expanded beyond the humble household kitchen garden, to incorporate a variety of agricultural activities comprised of a diverse range of community, governmental and non-governmental actors.

Conceptually, UA can be viewed as part of a broad urban food system that consists of a range of activities from production, distribution, retail and consumption of food and non-food products, as well as waste recycling and reuse. And similar to any system, be it economic or ecological, it has various integral components that, when healthy, maintain the performance of the system (Tisdell 2013). In other words, UA can be viewed as part of a complex urban food system that thrives with diversity—diversity in supportive and interconnected (as opposed to fragmented) policy spaces, for example, in health and nutrition, education, markets and commercial development, land use planning and zoning, provision of water and waste services and, of course, agriculture. If this diversity becomes compromised, then the system can struggle and eventually collapse (through lack of support). With this in mind, UA is unlikely to contribute to urban food security in the absence of connectivity in social, economic and environmental policy at national, state and local levels of government. In other words, the failure of a backyard garden in a South African township, to improve household food security, should not be used as a condemnation, or complete disregard, of UA as an ineffective and unsuitable concept for formal urban food policy consideration. UA is not a singular activity or practice that operates in isolated space. Its strengths are rooted in social, economic and environmental relationships and interconnectivity in, what are often fragmented, policy spaces (Mah and Thang 2013). It thrives where urban society and its institutions are connected, where overlap is identified linking community needs and government policy. Good urban governance, which is supportive of citizen mobility and encourages community-driven solutions to urban poverty and food insecurity, is as vital to UA as is biodiversity to soil.

This chapter established the purpose of this book and introduced urban governance as a framework for analysing challenges and opportunities for UA in the Global North and South, where experiences and outcomes in the social production of urban space for food security do vary. Given these differences and the inherently locally contextualised nature of UA, this book does not offer a 'blueprint' for UA and urban governance outcomes. Rather, it aims to present experiences in UA successes and struggles from cities in developed and developing countries, as a way to forward a discussion highlighting the importance of urban governance, particularly the vital role of city–community partnerships, for promoting and realising the social use value of UA and its related systems in creating more inclusive cities (as social and economic spaces), for all.

REFERENCES

Adger, W.N. 2000. Social and ecological resilience: Are they related? *Progress in Human Geography* 24 (3):347–364.

American Planning Association [APA]. 2010. Zoning for urban agriculture. In *Zoning for Practice March 2010*. Available at: https://www.planning.org/publications/publication/9006942/. Accessed 16 Apr 2018.

Barthel, S., J. Parker, and H. Ernstson. 2015. Food and green space in cities: A resilience lens on gardens and urban environmental movements. *Urban Studies* 52 (7): 1321–1338.

Battersby, J. 2011. Urban food insecurity in Cape Town, South Africa: An alternative approach to food access. *Development Southern Africa* 28 (4): 545–561.

Battersby, J. 2012. Beyond the food desert: Finding ways to speak about urban food security in South Africa. *Geografiska Annaler: Series B, Human Geography* 94 (2): 141–159.

Beall, J. 2002. Globalization and social exclusion in cities: Framing the debate with lessons from Africa and Asia. *Environment and Urbanization* 14 (1): 41–51.

Beall, J., O. Crankshaw, and S. Parnell. 2014. *Uniting a divided city: Governance and social exclusion in Johannesburg.* London: Routledge.

Beck, U. 1999. *World risk society.* Cambridge, UK: Polity Press.

Bek, D., T. Binns, and E. Nel. 2004. 'Catching the development train': Perspectives on 'top-down' and 'bottom-up' development in post-apartheid South Africa. *Progress in Development Studies* 4 (1): 22–46.

Brenner, N. 2012. What is critical urban theory? In *Cities for people, not for profit: Critical urban theory and the right to the city*, ed. N. Brenner, P. Marcuse, and M. Mayer, 11–23. New York: Routledge.

Brenner, N., and S. Elden. 2001. Henri Lefebvre in contexts: An introduction. *Antipode* 33 (5): 763–768.

Brenner, N., and N. Theodore. 2002. Cities and the geographies of "actually existing neoliberalism". *Antipode* 34 (3): 349–379.

Bush, R. 2010. Food riots: Poverty, power and protest. *Journal of Agrarian Change* 10 (1): 119–129.

Castells, M. 1983. *The city and the grassroots: A cross-cultural theory of urban social movements* (no. 7). Berkeley: University of California Press.

Colding, J., and S. Barthel. 2013. The potential of 'urban green commons' in the resilience building of cities. *Ecological Economics* 86: 156–166.

Crane, A., L. Viswanathan, and G. Whitelaw. 2013. Sustainability through intervention: A case study of guerrilla gardening in Kingston, Ontario. *Local Environment: The International Journal of Justice and Sustainability* 18 (1): 71–90.

Della Porta, D., and M. Diani. 2006. *Social movements: An introduction*, 2nd ed. Oxford: Blackwell.

DuPuis, E.M., and D. Goodman. 2005. Should we go "home" to eat?: Toward a reflexive politics of localism. *Journal of Rural Studies* 21 (3): 359–371.

Eckhardt, F., and Ingemar Elander. 2009. Urban governance: Introduction. In *Urban governance in Europe*, ed. F. Eckhardt and Ingemar Elander. Berlin: BWV Berliner Wissenschafts-Verlag.

Eizenberg, E. 2012. Actually existing commons: Three moments of space in community gardens in New York City. *Antipode* 44 (3): 764–782.

Elden, S. 2004. Between Marx and Heidegger: Politics, philosophy and Lefebvre's 'the Production of Space'. *Antipode* 36 (1): 86–105.

Feagan, R. 2007. The place of food: Mapping out the 'local' in local food systems. *Progress in Human Geography* 31 (1): 23–42.

Fisher, D.R., and W.R. Freudenburg. 2001. Ecological modernization and its critics: Assessing the past and looking toward the future. *Society and Natural Resources* 14: 701–709.

Folke, C. 2006. Resilience: The emergence of a perspective for social–ecological systems analyses. *Global Environmental Change* 16: 253–267.

Foster, J.B. 2002. *Ecology against capitalism*. New York: Monthly Review Press.

Food and Agricultural Organisation (FAO) of the United Nations. 2012. *Growing greener cities in Africa*. First status report on urban and peri-urban horticulture in Africa. Rome: FAO.

Food and Agricultural Organisation (FAO). 2013. *The state of food insecurity in the world 2013: The multiple dimensions of food security*. Rome: Food and Agriculture Organization of the United Nations.

Food and Agricultural Organisation (FAO) of the United Nations. 2015. *The state of food insecurity in the world (SOFI)*. Rome: Food and Agriculture Organization (FAO) of the United Nations.

Ghosh, J. 2010. The unnatural coupling: Food and global finance. *Journal of Agrarian Change* 10 (1): 72–86.

Guthman, J. 2008. Neoliberalism and the making of food politics in California. *Geoforum* 39 (3): 1171–1183.

Hajer, M.A. 1995. *The politics of environmental discourse: Ecological modernization and the policy process*, 40. Oxford: Clarendon Press.

Hammond, D. 2003. *The science of synthesis: Exploring the social implications of general systems theory*. Boulder: University Press of Colorado.

Harvey, D. 2003. *The new imperialism*. Oxford: Oxford University Press.

Harvey, D. 2008. The right to the city. *The New Left Review* 53 (Sept–Oct): 23–40.

Harvey, D. 2012. *Rebel cities: From the right to the city to the urban revolution*. London: Verso.

Hayden-Smith, R. 2014. *Sowing the seeds of victory: American gardening programs of World War 1*. Jefferson, NC: McFarland.

Healey, P. 2004. Creativity and urban governance. *disP-The Planning Review* 40 (158): 11–20.

Hodgson, K., M.C. Campbell, and M. Bailkey. 2011. *Urban agriculture: Growing healthy, sustainable places*. Washington, DC: APA Planning Advisory Service.

Kühn, M. 2003. Greenbelt and green heart: Separating and integrating landscapes in European city regions. *Landscape Urban Planning* 64 (1–2): 19–27. https://doi.org/10.1016/S0169-2046(02)00198-6.

Land Stewardship Project. 2010. How U.S. cities are using zoning to support urban agriculture. *LSP Factsheet* #21, 1–2. Available at www.landstewardship-project.org/repository/1/253/urbanagzoning.pdf. Accessed 24 Jan 2016.

Lawson, L.J. 2014. Garden for victory! The American victory garden campaign of World War II. In *Greening in the Red Zone: Disaster, resilience and community greening*, ed. K.G. Tidball and M. Krasny. Dordrecht: Springer.

Ledoux, T.F., and I. Vojnovic. 2013. Going outside the neighborhood: The shopping patterns and adaptations of disadvantaged consumers living in the lower eastside neighborhoods of Detroit, Michigan. *Health & Place* 19: 1–14.

Lefebvre, H. 1991. *The production of space*. Malden, MA: Blackwell.

Lefebvre, H. 1996 [1968]. *Writings on cities*. Oxford: Blackwell.

Lefebvre, H. 2003 [1970]. *The urban revolution*. Minneapolis: University of Minnesota Press.

Lindell, I. 2008. The multiple sites of urban governance: Insights from an African city. *Urban Studies* 45 (9): 1879–1901.

Mah, C.L., and H. Thang. 2013. Cultivating food connections: The Toronto food strategy and municipal deliberation on food. *International Planning Studies* 18 (1): 96–110.

Malan, N. 2015. Urban farmers and urban agriculture in Johannesburg: Responding to the food resilience strategy. *Agrekon* 54 (2): 51–75.

Marcuse, P. 2009. From critical urban theory to the right to the city. *City* 13 (2–3): 185–197.

McCarthy, J. 1997. Revitalization of the core city: The case of Detroit. *Cities* 14 (1): 1–11.

McClintock, N. 2014. Radical, reformist, and garden-variety neoliberal: Coming to terms with urban agriculture's contradictions. *Local Environment: The International Journal of Justice and Sustainability* 19 (2): 147–171.

Miller, C. 2003. In the sweat of our brow: Citizenship in American domestic practice during WWII-victory gardens. *The Journal of American Culture* 26 (3): 395–409.

Mitchell, D. 2003. *The right to the city: Social justice and the fight for public space*. Guilford Press.

Mittal, A. 2009. *The 2008 food price crisis: Rethinking food security policies, G-24 DP 56*. Geneva: UNCTAD.

Mougeot, L. 2000. Urban agriculture: Definition, presence, potentials and risks. In *Growing cities, growing food: Urban agriculture on the policy agenda*, ed. N. Bakker, M. Dubbeling, S. Gündel, U. Sabel-Koschella, H. de Zeeuw, and Zentralstelle fuer Ernaehrung und Landwirtschaft, 1–42. Feldafing: German Foundation for International Development.

Mun Bbun, T., and A. Thornton. 2013. A level playing field? Improving market availability and access for small-scale producers in Johannesburg, South Africa. *Applied Geography* 36 (1): 40–48.

Nally, D. 2011. The biopolitics of food provisioning. *Transactions of the Institute of British Geographers* 36 (1): 37–53.

Nicholls, W.J. 2008. The urban question revisited: The importance of cities for social movements. *International Journal of Urban and Regional Research* 32 (4): 841–859.

O'Hara, S.U., and S. Stagl. 2001. Global food markets and their local alternatives: A socio-ecological economic perspective. *Population and Environment* 22 (6): 533.

Okereke, C. 2007. *Global justice and neoliberal environmental governance: Ethics, sustainable development and international co-operation.* London: Routledge.

Pierre, J. 2011. *The politics of urban governance.* New York: Palgrave Macmillan.

Purcell, M. 2002. Excavating Lefebvre: The right to the city and its urban politics of the inhabitant. *Geojournal* 58 (2–3), 99–108.

Resnick, D. 2014. Urban governance and service delivery in African cities: The role of politics and policies. *Development Policy Review* 32 (s1).

Rogerson, C. 1992. Feeding Africa's cities: The role and potential for urban agriculture. *Africa Insight* 22: 229–234.

Roseland, M. 2012. *Toward sustainable communities: Solutions for citizens and their governments*, 4th ed. Gabriola Island, BC: New Society Publishers.

Rosin, C., P. Stock, and H. Campbell (eds.). 2013. *Food systems failure: The global food crisis and the future of agriculture.* London: Routledge.

Sanyal, B. 1985. Urban agriculture: Who cultivates and why? A case-study of Lusaka, Zambia. *Food and Nutrition Bulletin* 7: 15–24.

Sanyal, B. 1987. Urban cultivation amidst modernisation: How should we interpret it? *Journal of Planning Education and Research* 6: 187–207.

Säumel, I., I. Kotsyuk, M. Hölscher, C. Lenkereit, F. Weber, and I. Kowarik. 2012. How healthy is urban horticulture in high traffic areas? Trace metal concentrations in vegetable crops from plantings within inner city neighbourhoods in Berlin, Germany. *Environmental Pollution* 165: 124–132.

Sidaway, J.D. 2000. Postcolonial geographies: An exploratory essay. *Progress in Human Geography* 24 (4): 591–612.

Smith, D., and D. Tevera. 1997. Socio-economic context for the householder of urban agriculture in Harare, Zimbabwe. *Geographical Journal of Zimbabwe* 28: 25–38.

Soja, E. 2010. Spatialising the urban. Part 1. *City: Analysis of Urban Trends, Culture, Theory, Policy, Action* 14 (6): 629–635.

Thaman, Randy. 1975. *Urban gardening in Papua, New Guinea and Fiji: Present status and implications for urban land use planning.* Suva: University of the South Pacific.

Thornton, A. 2008. Beyond the metropolis: Small town case studies of urban and peri-urban agriculture in South Africa. *Urban Forum* 19 (3): 243–262.

Thornton, A. 2012. *Urban agriculture in South Africa: A study of the Eastern Cape Province.* Lewiston, NY: Edwin Mellen Press.

Thornton, A., K. Lyons, and S. Sharpe. 2018. Community gardens for social and food justice: The case of urban agriculture in Australian cities. In *The Routledge handbook of community development research*, ed. L. Shevellar and P. Westoby. London: Routledge.

Thornton, A., J. Momoh, and P. Tengbe. 2012. Institutional capacity building for urban agriculture research using participatory GIS in a post-conflict context: A case study of Sierra Leone. *Australasian Review of African Studies* 33 (1): 165–176.

Tidball, K.G., and M. Krasny (eds.). 2014. *Greening in the red zone: Disaster, resilience and community greening*. Dordrecht: Springer.

Tilly, C. 1999. From interactions to outcomes in social movements. In *How social movements matter*, ed. M. Giugni, D. McAdams, and C. Tilly. Minneapolis: University of Minnesota Press.

Tisdell, C.A. 2013. *Competition, diversity and economic performance*. Cheltenham: Edward Elgar Publishing.

UN-Habitat. 2015. Governance. Available at: http://unhabitat.org/urban-themes/governance/. Accessed 20 Oct 2017.

Von Bertalanffy, L., and A. Rapoport (eds.). 1963. *General systems. General systems yearbook*. Ann Arbor: Society for General Systems Research.

Walker, R.E., C.R. Keane, and J.G. Burke. 2010. Disparities and access to healthy food in the United States: A review of food deserts literature. *Health & Place* 16 (5): 876–884.

Warshawsky, D. 2014. Civil society and urban food insecurity: Analyzing the roles of local food organizations in Johannesburg. *Urban Geography* 35 (1): 109–132.

Wittman, H., A.A. Desmarais, and N. Wiebe. 2010. The origins and potential of food sovereignty. In *Food sovereignty: Reconnecting food, nature and community*, ed. H. Wittman, A.A. Desmarais, and N. Wiebe, 1–14. Oakland, CA: Food First.

CHAPTER 2

Urban Agriculture: Overview of the Field and Early Models of Urban Food Governance

Abstract This chapter will present an overview of the field Urban Agriculture (UA) scholarship from the global North and South. Using an urban governance framework, this overview will focus on the relationship between UA and food security and the use of urban space for UA activities, from earlier periods of need, to the contemporary era of a globalising urban population, and to what extent UA is a solution to, or is symptomatic of, urban decline and a misunderstood fact of urban life. These issues are intertwined and pertinent, as social movements for urban food activism are clearly increasing in cities, globally, with multi-scalar impacts and policy outcomes that indicate the connected and disconnected nature of urban dwellers and their institutions.

Keywords Alternative food networks · Urban agriculture · Food security · Food policy · Victory gardens · War gardens · 'rights to the city' · Social use value · Sustainable Development Goals

Since the 1980s, academic interest in urban agriculture (UA) has grown, beginning with debates on its general concepts regarding its definition, to applied studies not limited to production systems, urban farmers, land tenure, production phases and market access (Sanyal 1987; Rogerson 1993; Mougeot 2000; Thornton 2008, 2012; Nel et al. 2009). UA can be understood as the cultivation, processing, marketing and distribution of food, forestry, horticultural and aquaculture products that occur

© The Author(s) 2018
A. Thornton, *Space and Food in the City*,
https://doi.org/10.1007/978-3-319-89324-2_2

19

in built-up 'intra-urban' areas (Mougeot 2000). It is also a holistic system that goes beyond food production, as it has the potential to close the food-waste loop in urban areas and contribute to low-carbon urban futures (Hopkins 2008). Interest in UPA is increasing as a result of price fluctuations for basic food staples, inequalities in agro-food networks (Thornton 2011) and persistent poverty, particularly in sub-Saharan Africa (Kneafsey 2010; Larder et al. 2014; Lynch et al. 2012; Thornton 2012).

GLOBAL OVERVIEW OF URBAN AGRICULTURE

UA is not a new phenomenon, noting its existence in pre-industrial societies, with roots in ancient civilisations worldwide (Mougeot 1994; Nugent 2000; Van Veenhuizen et al. 2001). The 'Hanging Gardens of Babylon' built about 2500 years ago probably were humankind's first example of UA. Further, several ancient civilisations had developed complex UA systems and technologies; for example, Persian and Roman sites created advanced hydraulic facilities and agricultural drainage schemes, respectively, and 'the Islamic empire' used its 'postal service' to gather information on food prices and food supply to prevent shortages (Mougeot 1994). North America's Mississippian cultures (1050–1250) were supported by intensive riverine horticulture (Mougeot 1994). However, as technology ushered urban human settlements into the industrial era, urban farming practices were deemed inappropriate and were subsequently assigned to rural regions. The more recent history of urbanism, associated with the industrial revolution, has, for many Western countries, resulted with the separation of 'urban' from 'agriculture'—except for recreational gardening or in times of crisis. As discussed in Thornton (2012: 4), the importance of UA for survival and earning a livelihood among the urban poor was evident from earlier field studies undertaken by French geographers in West Africa during the 1950s (Mougeot 1999), and accounts of its practice in the South Pacific islands in the 1970s (Thaman 1975).

Discussed in a later section below, perhaps the most convincing argument for nationwide urban agricultural land use planning in global North cities, are the 'War Gardens' and 'Victory Gardens' of the First and Second World Wars. These were exceptional in the level of multi-scalar institutional involvement, from the community, city, state and region, and involving stakeholders from households, private and public sectors. Although the post-war economic boom led to the demise of this form of coordinated land use, social activism in US cities in the 1960s–1970s saw the re-emergence of community gardens (Henderson and Hartsfield 2009). However, it was not until the late 1970s and 1980s that UA shifted "from being a scientific curiosity, to being an urban policy issue and development tool" (Mougeot 1999, in Thornton 2012: 5). UA has increasingly become synonymous with sustainable urban development since the 'Brundtland report' popularised the importance of sustainable development (Thornton 2012: 5). The Brundtland report urged governments in the developing world to consider UA as an important component of urban development in "provid[ing] more [urban] green space", and food security for the urban poor (Bruntland 1987: 254). In diverse urban settings globally, UA is rigorously pursued, particularly at the grassroots level (and to some extent at the formal urban policy level), as one response to the pressures of an expanding global urban population, escalating urban poverty and inequality in the global food system (Thornton 2017).

In the 1980s and 1990s, a steady stream of optimistic case studies of UA (Smit et al. 1996, 2001) was criticised as overstating its real productive impact (Ellis and Sumberg 1998). These case studies were determined as too subjective, with findings based on limited qualitative and anecdotal statements advocating for UA, as a panacea for food security and poverty concerns (Zezza and Tasciotti 2010; Badami and Ramankutty 2015).

Discussed in Thornton (2012: 5), UA from the 2000s was described as a significant urban food production and supply system, often using complex, capital-intensive methods (Yi-Zhang and Zhangen 2000). Moreover, The World Resources Institute (2000: 144) had reported that "in Kenya and Tanzania, two in three urban families use urban agriculture", while in Taiwan "more than half of all urban families are members of farming associations". Despite these claims, UA in some parts of Africa was seen as less prolific (Thornton 2012: 5). In urban areas in countries such as Uganda (Maxwell 1994) and Zimbabwe (Mbiba 2000), and to a certain extent Ghana (Armar-Klemesu and Maxwell 2000), UA appeared to have had a limited impact on the livelihoods of the urban poor, while outside of the African continent, for example, in Bangkok, Madrid and San Jose, CA, up to 60% of the metropolitan area was described as under some form of cultivation (Thornton

2012: 5). Literature highlighting significant challenges to UA also increased in the last couple of decades, as urbanizing populations, in the North and South, were connected to loss of fertile urban and peri-urban land to housing development (Nel et al. 2009; Thornton et al. 2012; Thornton 2017). In the last decade, many observers have called for more quantitative assessments into the food security benefits of UA, particularly in sub-Saharan Africa, discussed below—a region that missed several MDG targets for eradicating hunger and malnutrition (Zezza and Tasciotti 2010; Martellozzo et al. 2014).

Urban Agriculture in the 'Global South'

This section will briefly outline some of the key urban challenges facing many local governments, urban planners and policymakers in the developing world, who seek solutions for sustainable urban development. These key challenges include urbanisation, poverty alleviation, food security and economic crises. These factors have been widely discussed in the literature as modern problems that can potentially be addressed by encouraging, requiring and supporting (at multiple scales) the practice and expansion of an 'age-old practice', namely UA.

Urbanisation

The United Nations estimates that between 2007 and 2008, the global population became more urban than rural, for the first time in the history of human settlements. In 2014, 54% of the world's population lives in an urban area. Many observers claim that a greater part of population increases will occur in developing countries, with 56% of Africa's population projected to be urban by 2050—and the urban poor to suffer the most (Borel-Saladin 2017). As developing states struggle to adjust to challenges associated with urbanisation, such as housing provision and unemployment, the urban poor have largely been left to their own devices (in the absence of urban food policy solutions) to cope with inadequacies in the urban food supply (Nel et al. 2009; Thornton et al. 2012). However, to a certain extent some governments are promoting self-reliance in food provision (with varying levels of technical, financial and policy support) through UA (Companioni et al. 2002; Krishnan et al. 2016). As Thornton (2012: 12) argues, the role of official or government support of UA may facilitate production efficiencies in delivering higher yields, improving labour productivity and value adding (Yi-Zhang and Zhangen 2000; Orsini et al. 2013).

Poverty Alleviation

A crucial aspect of UA for many observers is that the poorest households receive, at a minimum, subsistence level benefits from its practice. This perception, among others (discussed in the Berlin case study, below), may effectively prevent UA debates from moving forward. Earlier studies by Smith and Tevera (1997: 25) had argued, "there is a lack of distinction made between [urban agriculture] as a survival strategy for the poor and as a viable commercial activity, whose proponents are seeking to take advantage of the market opportunities afforded by rapid pop[ulation] growth, or by expanding markets". Other factors, beyond traditional, could exist that determine how significant a home garden plot or livestock is to a household. These factors could include variation in the size and location of this activity, how it is applied, as well as its use, which is not necessarily determined by income. Moreover, different types of individuals, groups, enterprises, institutions and governments typically determine how UA is structured; and income levels can condition land access, tenure security and labour, which in turn affect production, the mix of inputs, volumes and distribution (Thornton 2012).

Food Security

The FAO (2015) defines food security as access by all people at all times to the food required for a healthy life; at the household level, at issue is the household's ability to secure enough food to ensure adequate dietary intake for all its members. Localising food production is increasingly viewed as one way to improve access and availability of nutritious low-cost food for low-income urban households (Thornton 2012: 13). In depressed urban areas, the losses of such coping mechanisms (subsistence production, community networks of sharing) are compounded by the shortfalls of "modernization and monetization" (Sahn 1989, in Thornton 2012: 13). As Thornton (2012: 13) argues, this has been the case in developing countries, where dependency on a cash economy has shifted the urban poor away from an (agri)culture of home cultivation. Through supporting the practice and expansion of UA, governmental and non-governmental policymakers can limit the dependency on the cash economy by the urban poor, in particular in the event of 'seasonal undulations' in supply and cost of urban food (Thornton 2012: 13).

With African cities projected to host nearly 25% of the global urban population by 2050 (Cobbinah et al. 2015), identifying pathways to strengthen urban food security is a key development challenge (Crush and Frayne 2011; Tawodzera 2011; Battersby 2012). Discussed in Thornton (2018), no other region had experienced more urban food riots than African countries during 2007–2008 food crisis, and again in 2010–2011, which led to greater awareness of causal factors driving food protests (Sneyd et al. 2013; Bohstedt 2014). Of interest to this book, studies on urban food riots in this region found that citizens mobilized to demonstrate a profound dissatisfaction and lack of "fairness in rights to food, its distribution, and accountability for food supply failures" (Hossain 2009: 2). What is not happening in the region, is citizen mobility evolving into the type organised urban activism (see Chapter 3) that seeks engagement with city officials (and other food system stakeholders) in negotiating city–community strategies to strengthen the urban food supply or local food systems. UA is seen as a stimulus for political and social stability, as it is an activity mobilising the urban poor in a 'moral economy' to ensure food fairness in the North and South, for example, in Cuba, Sierra Leone and Detroit (Thornton 2018). Recent reviews (Poulsen et al. 2015: 138) of the case study literature find the UA substantially contributes to farming households' food availability in some settings, and the introduction (or improvement) of favourable city planning policies may diminish barriers to urban food production.

The post-2015 Sustainable Development Goals (SDG) emphasize ending hunger, achieving food security for all and promoting sustainable agriculture (SDG2), while creating inclusive, safe, resilient and sustainable cities (SDG11) (United Nations Development Program—UNDP 2015a: n.p.). Discussed in Thornton (2018), urban food security is curiously missing from SDG11 (the 'urban SDG'). However, as Thornton explains, there appears to be an element of 'co-dependency' in the SDGs, where "achieving the targets under SDG11 sets the stage for achieving targets in many of the other SDG goals" (UNDP 2015b: n.p.). Related to this, the widely cited definition of food security from the FAO (2015: n.p.) is purposefully inclusive, stating that food security exists when all people, at all times, have physical, social and economic access to sufficient, safe and nutritious food that meets their dietary needs and food preferences for an active and healthy life.

Economic Crises

UA has thrived in in countries experiencing economic crises, such as Cuba (Rosset and Benjamin 1994; Endres and Endres 2009) and Indonesia (Purnomohadi 2000), as well those undergoing structural adjustment (Simon 1995; Thornton 2012). The impact of structural adjustment programs (SAP), or austerity measures, severely affected urban food consumers and rural producers. Discussed in Thornton (2012: 13), SAPs required borrowing countries to liberalise their markets through decentralisation and privatisation processes. This included cuts in social safety-nets, such as food subsidies and transport of food produced on rural farms to consumers in urban areas (Islam and Nag 2010), in order to facilitate the servicing of foreign debt (Thornton 2012). A reduction of public expenditures often translated into the termination of domestic consumer food subsidies. The loss of these subsidies directly added a significant burden to the urban in meeting their food and nutritional needs (Maxwell et al. 2000). Although the following claim is heavily debated (Frayne et al. 2010; Crush et al. 2011), urban and peri-urban agriculture is contributing to urban food security in many African cities, despite poor urban governance and restrictive policy environments (such as bylaws and ordinances), for example, in Cameroon (Rogerson 2003); Guinea-Bissau (Lourenco-Lindell 1997); Sierra Leone (Lynch et al. 2012; Thornton et al. 2012); Tanzania (Wenban-Smith et al. 2016); Zambia (Nel et al. 2009; Thornton et al. 2010) and Zimbabwe (Mbiba 2000; Ncube and Ncube 2016).

Urban Agriculture in the Global North

UA is increasingly playing a role in the development of local and 'alternative' food networks (AFN). The United States Department of Agriculture defines local food as food transported less than 400 miles from where it had originated (Martinez et al. 2010), compared to the conventional global distribution system with food traveling up to 3000 miles (in the United States, 1500 miles on average), thus requiring added packaging and refrigeration to maintain government food quality standards. Smith and McKinnon (2007), among others, popularised the health and environmental benefits of eating locally in their book, *The 100-mile diet: A year of local eating (Random House, Canada)*. Smith and McKinnon set out to meet their food needs from sources 100 miles from their Vancouver apartment, as the title suggests, for one year. Urban and peri-urban food producers from larger urban farms in the metro region and smaller inner-city community and backyard gardens contributed to this effort.

Overall, UA has re-emerged in many parts of the western world as an integral part of an urban food strategy (Oosterveer and Sonnenfeld 2012). I say 're-emerged', because we have been down this road before, during the First and Second World War eras, where domestic agriculture—in urban and rural areas—was promoted and developed at a large scale to strengthen food security during crisis periods. An important aspect of this was strengthening food knowledge (a vital dimension of food security) in urban areas for adults and children. The following section explores national food garden programs, before, during and after the world war periods, in more detail.

War Gardens and Victory Gardens

Although the practice of growing food in and around urban settlements has a long history, the notion of communal and community gardens in 'modern' cities can be traced back to the allotment garden system in the United Kingdom, which the Small Holdings and Allotments Act 1907 required local authorities to provide "where demand existed" (Acton 2011: 49). There was an increase in allotment demand during WWI, as people were encouraged by the UK Government's war-time propaganda campaigns, to grow food in response to food shortages. This included food growing in towns and cities throughout the country. At its peak, there were 1.5 million urban garden allotments during World War I and 1.75 million in World War II.

During the war, the government's promotion of urban food gardening was evident, coupled with food rationing. At the outset of WW1, the UK imported 60% of its food supply, which was summarily disrupted by German U-boat campaigns. The Minister of Food Control during WWI, followed by the Minister of Food in WWII, to oversee the rationing of food (Hayden-Smith 2014: 149). By 1916, food prices in the UK had tripled and long queues for sugar, meat and staples such as potatoes were the norm (Weiss 2008: 10). To increase domestic food supply, the UK Government largely focussed on improving food production efforts on rural farms. As the war created a shortage in male farm labour, the Women's Land Army (WLA) and Women's Institutes (WI) came to fruition, much to the initial disapproval of male farmers and initial indifference of the UK Government (Adie 2013: 259–264; Weiss 2008: 8–9). The WLA was government funded, under the auspices of the new Women's Branch of the Food Production Department (Weiss 2008: 11). The WLA was comprised of an Agriculture, Forage and Timber Section, of which women enlistees could choose to serve. Before they could serve on the farms, women enlistees undertook intensive farming training programs overseen by various national and district level organisations, such as the Women's National Land Service Corps, Women's War Agricultural Committee (WACs), Women's Farm & Garden Union. By 1916, although official figures are disputed, 57,500 women (rural and urban) had registered for agricultural work and 29,000 were working on farms and dairies (Weiss 2008: 11). With the end of the war, urban allotments were requisitioned mainly for the purposes of economic development, despite continuing demand for allotment space (Acton 2011: 49). Although the WLA was disbanded by 1920, the concept returned to service through the 'Dig for Victory' campaign during Second World War (WWII).

United States

In the United States, urban growing food during times of need preceded the First and Second World Wars. In the nineteenth century, food shortages in cities, notably in 1837, 1863 and 1893, triggered major food riots. In 1893, the Mayor of Detroit launched the 'Detroit Experiment', popularly called 'Potato Patch Farms', to help alleviate urban food insecurity. Particularly effective during the two world wars, was tapping into the American sense of rugged individualism and self-reliance. These are bound to the notion of America's origins as a nation of farmers (Hayden-Smith 2014: 31). This 'rural past', it was believed, could influence the growing urban population to take to the soil (ibid.: 8). As WWI disrupted the global food supply, an immense and determined effort, at multiple scales, was put into motion to produce enough food for the domestic supply and for the war effort. The involvement of institutions, at federal, state non-governmental levels, struck a balance between American individualism and voluntarism as a patriotic act. Through linking a culturally embedded ideal of the productive 'frontier pioneer' to address national food shortages, it was possible to motivate urban dwellers (as well as rural) to significantly increase local food production during times of crisis.

For cities in western countries, particularly in the UK, United States and Canada (in Australia as well, though not as effective, see McKernan 1983: Chapter 6), the establishment of 'Liberty/War Gardens' and 'Victory Gardens' during the First and Second World Wars, were critical to strengthen national food security as commercial production was directed abroad to feed allied nations and troops (Pack 1917; Bourne 1942; Lawson 2005). The importance of urban and suburban food production arose during the First World War (WWI), driven by the National Emergency Food Garden Commission (NEFGC), National War Garden Commission (NWGC) and the United States School Garden Army (USSGA). From a nationwide 1917 NEFGC survey, nearly 3 million urban and suburban food gardens, "where none had grown before", had contributed to the nation's food supply, and 'conservatively' valued at USD350,000,000 (Pack 1917: 203). It also reintroduced the practice of home canning and drying of fruits and vegetables "an art almost forgotten since our grandmothers' days" (Pack 1917: 203). At the end of the war, the NEFGC had called for any remaining 'slacker land', in urban and suburban areas, to be converted into 'Victory Gardens' (a term used in the Second World War), as export food production from the United States was in heavy demand. At the close of the WWI, US agricultural production was needed in several European allied countries, as well as oppressed and neutral nations, feeding an aggregate of 290 million people (Pack 1919: 2).

Understanding the significance and impact of the 'War Gardens' and 'Victory Gardens' on food security is important, from an institutional and food policy perspective, as the war-time food gardens represent an ambitious effort, at multiple scales, to respond to a global crisis that had threatened the global food system, by galvanising the population behind a nationalistic productive ethos to shorten the food supply chain. National garden campaigns during the war periods offer valuable lessons in the context of strengthening urban resiliency in an era of global warming, as it affects the global food system and long food supply chains.

URBAN FOOD GOVERNANCE IN THE TWENTY-FIRST CENTURY

A common thread held in cities, globally, is the conflict between the sociocultural need for public social spaces and neoliberal market demands for urban commercial and residential development. It is in this context that local networks of small-scale inner-urban and larger scale peri-urban food production (or UA), in green zones and rooftops, in suburbs and in smaller towns, must be recognised as contributing to the food needs of its urban populations. In parts of Europe and North America, for example, these networks exist and are expanding under community supported agriculture and other AFN (Goodman et al. 2011). Such methods are proving to be viable alternatives to mainstream monoculture crop production (see IAASTD 2009), with permaculture methods used in urban areas where space to grow food is limited, particularly in the inner-city regions, and often requiring more intensive production. Discussed in a later section, city managers are reviewing bylaws and ordinances in an effort to zone for various form of UA. However, this is proving more effective in cities in the global North, less so in the South. Among the important changes to city ordinances has been commercial zoning for UA, thus allowing for the production and marketing of urban produce and foodstuffs sourced from backyards, community and market gardens and urban farms. The following section will explore the impact of local level institutions, or grassroots groups, as drivers of social movements in claiming their 'rights to the city' for local or urban food production.

Social Movements and the Production of Food Space

Urban spaces are becoming increasingly commodified and, in response, communities are claiming their 'rights to the city'. A 'right to the city' is essentially a collective response by urban communities to 'take back' neoliberal dominated urban space for social use (Lefebvre 1991). For David Harvey (2008: 23), rights to the city is also the "freedom to make and remake our cities and ourselves [and is] one of the most precious yet most neglected of our human rights". This response is reflected in Gibson-Grahams' (2008) diverse economy approach, where people engage in alternative activities that may exist alongside, or challenge, the dominant, mainstream neoliberal economy (or capitalist system).

Conceptualizing UA as mechanism or 'system' through which people assert their urban 'rights', as an alternative social production of space, is an emerging area for urban theorists and critical geographers in exploring city–community relations and its theoretical implications. For example, Eizenberg (2012: 778) explored community gardens in New York City as 'actually existing commons', where city-wide coalitions engaged in dialogue with city officials and the wider public, to identify strategies for protecting and maintaining over 400 community gardens (for example, non-profit expropriation of land from the market as public land trusts). Studies from San Francisco, Stockholm (Barthel et al. 2015: 1329) and Berlin (Colding and Barthel 2013: 159) suggest that better informed decisions in the collective use and management of urban space for community gardens are possible when groups of civic actors organise, as coalitions or social movements, to negotiate with powerful interests for policies to accommodate urban food production. These studies make important contributions in applying critical urban geographies to explore particular aspects of UA, which offers useful framing for a discussion of the possibilities for city–community partnerships for sustainable cities. Whereas Lefebvre's theorizing does not exclude the possibility of city–community partnerships (Purcell 2002), the research discussed in this book seeks to explore urban theoretical and applied knowledge on these possibilities as urban space is contested by competing land users amid population pressures and economic development.

Alternative Food Networks and Urban Governance

Urban governance is a broad concept that includes the role of institutions and individuals in creating an enabling environment, such as community–city partnerships, to effectively respond to the needs of all urban residents (UN-Habitat 2015). Urban social movements are a grassroots or community-led response to local concerns of social equity, environmental and food justice themes. Social movements are often influential agents in shaping good urban governance outcomes. These outcomes include improved interconnectivity among municipal, community and private stakeholders in the urban food system and re-prioritising policies that improve urban resilience, social cohesion and local and alternative food economies, in the face of multiple urban sustainability challenges. Alongside rapid urban population growth, various forms of UA activities are notably expanding globally.

The connection between AFN and urban governance lies in its focus on how people engage with and enact on their understandings of 'local' and 'alternative' food production activities. In cities in North America and Europe, in particular, it appears that community engagement in local and alternative urban food systems seek to overcome an often 'antagonistic relationship' (Brenner 2012: 19) that exists between socially just spaces and actually existing neoliberal spaces, as reconciling this relationship encompasses broader concerns regarding collective urban rights. This view embraces Harvey's (2008: 23) assertion that "the freedom to make and re-make cities and ourselves is [...] a neglected human right". In this statement, Harvey emphasizes Lefebvre's thoughts on the potential power of collective urban rights to change cities. For Lefebvre (1996 [1968]), every society produces its own space, which can have 'non-productive' use value that is in constant conflict with the neoliberal spatial production (Elden 2004). This alienation reflects the role of power relations in spatial production and the struggle for spatial justice or 'rights to the city'. As Thornton (2017) argues, these issues are relevant to radical urban theorists in discourses on alternative production of space and alternative and diverse economic geographies (Leyshon et al. 2003; Gibson-Graham 2008).

In many western countries, AFN exist as a collection of production and marketing activities at regional and local levels, as seen in farmers' markets, food-box and food swap initiatives, community farms and market gardens. The emergence of AFNs is a type of localised response to the social, economic and environmental impacts of the global industrial agro-food system. Initially viewed as a mechanism to preserve rural agri-based livelihoods in the countryside, AFNs are also understood, in both rural and urban contexts, as a non-conventional or non-corporate attempt to reconfigure relationships between food producers and food consumers (Venn et al. 2006). There is a wealth of knowledge in the rural sociology literature on alternative agro-food networks as an outcome of a 'quality-turn' in food preferences and desires (Goodman 2003; Marsden et al. 2000; Maye et al. 2007). These 'preferences and desires' for certain foods has drawn criticism of AFNs as an exclusive system comprised of affluent consumers of privileged 'foodie' lifestyles (Donald and Blay-Palmer 2006; Guthman 2008). However, the global rise of AFNs indicates broader concerns among consumers over the exploitative labour and marketing processes in the global industrial food system (Barnett et al. 2005; DuPuis and Goodman 2005). In other words, consumers increasingly want to know where their food comes from (small farm or commercial feedlot) and how was it produced (organic, GMO). In terms of farm animal welfare (FAW), consumers want to know under what ethical standards were livestock reared and processed (slaughtered) for the market (Lever and Evans 2017).

As discussed by Larder et al. (2012: 4), localising food has become associated with the AFN, with an aim to foster social and environmental equity in food production-distribution-consumption. This is seen as a response to the negative impacts of the neoliberal globalised food system, understood as the persistence of chronic hunger and high costs of food associated with industrialised food production, complex marketing and long distribution chains (Larder et al. 2012). Gibson-Graham (2008) provide a critical framework that can assist in assessing the emergence of alternative food economies (spatially tied or localised producers, consumers and markets) and local food choice behaviour. This can be useful in assessing social and economic value of UA activities, such as community gardens, as a contributor to AFN (Thornton 2017). In other words, it can help explain relationships in the local food economy that link local production to local consumption, such as farmers' markets and community gardens (Thornton 2017).

Gibson-Graham's (2008) diverse economies approach can also explain the existence of alternative economies, beyond typical analyses of mainstream neoliberal economic activities (Thornton 2017). Analyses of these activities are located in the top row of the diverse economies framework (DEF) (Table 2.1). Each column, in Table 2.1, reveals all other alternative and non-formal elements that make up the diverse economy. When considering the role and impact of AFN in urban governance, diverse economies can improve relational understandings (as diverse economies will differ from place to place) of 'alternative' and 'conventional' food systems from the point of view of 'actually existing' community gardeners. This approach is useful when exploring why city or municipal level support for alternative and local food systems can be "sporadic

Table 2.1 Representation of the Diverse Economies Framework (DEF)

Transactions	Labour	Enterprise
MARKET	WAGE	CAPITALIST
Alternative market	Alternative paid	Alternative capitalist
Underground market	Self-employed	State enterprise
Co-op exchange	Cooperative	Green capitalist
Barter	Reciprocal, In kind	Socially responsible firm
Informal market	Work for welfare	Non-profit
Non-market	Unpaid	Non-capitalist
Household flows	Housework	Communal
Gift giving	Family care	Independent
Indigenous exchange	Neighbourhood work	Feudal
Hunting, gathering	Volunteer	

Source Adapted from Gibson-Graham (2008)

and certainly not voiced in policy circles" when comparing cities within a particular country, as well as when making international comparisons (Maye and Kirwan 2010: 5).

Discussed in Thornton (2017), possible outcomes of a diverse economies approach to alternative food economies include determining the relevance of local AFN to community gardens and benefits and reasons for civic engagement in urban food production. More broadly, alternative social production of urban space can offer contributions, theoretically and empirically, to the wider research field on the dynamics of alternative food economies and the emergence of local responses to the negative impacts of the global food system (Thornton 2017). The Australian case study, below, will place emphasis on the social benefits of local urban food activism, as 'non-monetary' drivers of a diverse—and localised—food network (Jarosz 2009; Maye and Kirwan 2010; Low and Vogel 2011; Martin 2011).

Mainstreaming of 'Alternative' Agri-Food Initiatives

As supermarkets, chain stores and governments have become increasingly visible in social and sustainable food issues, the 'alternativeness' of agri-food initiatives is being questioned by academics and 'food activists' as reflecting the neoliberal notion that social change is best pursued through the market (Alkon 2013). When viewing AFNs critically, to what extent are they proposing 'alternatives' to the negative impacts of the corporate agro-food system? Are AFNs reproducing and reinforcing 'actually existing' neoliberalism (Lyson 2004; Maye and Kirwan 2010). This issue of the potential for normalising 'alternative' local initiatives through their incorporation as a neoliberal project is discussed below.

Zoning for UA: Is It Still Alternative, or Is It Normalising? Does It Matter?

In the United States, several cities have tangible urban agricultural systems in place, with the cities of Boston, Los Angeles, San Francisco, San Antonio, Pittsburgh, Milwaukee, Seattle, Portland and, recently, Detroit currently zoning for urban agricultural space. In Detroit, a city experiencing rapid urban decline in the wake of the global financial crisis, it is estimated that 5000 acres under tillage could provide up to 28,000 jobs and 70% of the city's food needs (Detroit Free Press 2011).

For the City of Boston, the Conservation Law Foundation (2012) claims UA could reduce the city's greenhouse gas (GHG) emissions; with 50 acres (over 20 hectares) of properly managed soils sequestering about 114 tons of carbon dioxide (CO_2) per year, and potentially enabling an additional CO_2 reduction of up to 4700 tons per year, and generating approximately 1.5 million pounds of fresh produce for sale into local markets. Many of these cities in North America have passed legislation for UA that ensures its legal status, despite significant urban population densities.

In December 2013, the City of Boston established 'Article 89', which essentially legalised, through a two-year zoning and community consolation process, various forms of UA, including rooftop gardens, in the inner city and wider metropolitan area. This process included consultation with local communities and an urban farm pilot project. In Boston, community groups and city planners emphasised that UA is a participatory, proactive response from the city to address issues of social, economic and environmental and food justice (Thornton 2017).

In line with the global view of UA, food production and marketing had long existed in the city (Warner 1987), but no frameworks had previously existed to support it (Kaufman and Read 2016). As Popovitch (2014: 1) explains, "with over 40 food truck companies, a pilot residential composting program, 200 community gardens, 100 school gardens and 28 farmer's markets, Boston was in need of a framework for its growing sustainability efforts". In one example, the non-profit *Victory Programs* operates an urban farm (*ReVision Urban Farm*) on land that was previously vacant city land. 'ReVision' hopes to address issues of food and social justice in the neighbourhood of Dorchester, where convenient access to fresh food outlets are extremely limited. The lack of availability of fresh produce creates a habituated reliance on foods with low nutritional value, such as highly processed foods and fast foods. Upon determining that local residents do not "understand vegetables" and therefore leafy greens do not feature at the dinner table, *Victory Programs* started teaching cooking skills to residents ('Rachel', *ReVision* Urban Farm, personal communication, 17 April 2013). Many of these residents eventually participate in the urban farming activities, such as developing seedlings and selling produce at local farmer's markets and community supported agriculture. Insights learned from the passage of 'Article 89' and ReVision Urban Farm reflects Foo et al. (2014: 176), in their study of vacant land in Boston. Foo et al. in recognised that community gardening is an "activity that challenges negative perceptions of economically depressed areas by creating new functions and values of space within neighborhoods". They found that policy change in urban land use begins at the neighbourhood-scale, where residents can influence urban land use policy and transform vacant city land into meaningful spaces of social interaction.

Urban Food Governance in African Cities?

As a point of contrast cities in the 'South' appear steadfastly in support of post-WWII ideas of modernisation through urban industrialisation for economic growth (Thornton et al. 2010). Since the 1990s, and particularly since the World Conference on Human Rights (1993) and World Summit for Social Development (1995), the development paradigm has shifted towards more participatory and inclusive approaches that emphasise social development, as central to development and an equal partner to economic concerns (Desai 2014). This view has contributed to dramatic transitions in many cities in the developed 'North', which emphasise varying levels of commitment to urban social equity in planning decisions, especially for spatially marginalised communities. In these communities, social equity often includes achieving food security and 'food justice', which is expressed through various forms of UA (Thornton et al. 2018). By contrast, in cities in the 'South', the scale of inequality due, in large part, to differentiated impacts of successive development strategies, does not easily lend itself to a convergence in ideas and action for re-conceptualising the purpose, potential and promise of inclusive, citizen-led urban spatial (re)production. Although calls for more inclusive urban governance are increasing in parts of Latin America and South Africa (Brown 2013; South African Government 2012; Huchzermeyer 2012, 2014; Thornton and Rogerson 2013), such calls struggle for recognition elsewhere. As the following example of Lusaka, Zambia, reveals, UA is officially viewed as a practice that is out of place in 'the modern city', despite the fact that it is a fixture in that city, albeit, on an informal basis (Rakodi 1988; Nel et al. 2009; Thornton et al. 2010).

Lusaka, Zambia

Previous research has revealed that Lusaka's city planners believe that UA 'is not modern' and urban spaces need to be reserved for investment opportunities (Thornton et al. 2010). This is despite the fact that institutional support for UA in Zambia is not completely unprecedented. From the 1960s to the 1970s, for example, there were national plans to encourage and support urban gardening, including upgrading squatter settlements by increasing plot sizes to allow for food production (Rakodi 1988). In Zambia's 'Third National Development Plan' (1978–1983), former president Dr. Kenneth Kaunda specified the need for making 'self-sustaining cities' by increasing urban food production (Rakodi 1988; Thornton et al. 2010). Kaunda's humanistic approach of combining self-sufficiency in agriculture with import substitution did not, however, accord with the aspirations of the more neoliberal-bent members of his government, who eventually accepted the World Bank's deterministic 'modernisation by industrialisation' prescriptions (Quick 1977). Ultimately, Kaunda's vision for 'self-sustaining cities' in Zambia never materialised and UA remained a marginalised and informal activity, reflecting the situation in other African cities (Thornton et al. 2010).

In contrast, the Johannesburg Development Agency (2015) boasts of allocating urban space, previously earmarked for development, for 42 urban gardens, as well as inner-city plots and rooftop gardens in city neighbourhoods including Hillbrow, Joubert Park, Troyeville, Newtown and Fordsburg. There is also institutional-level interest to include informal peri-urban farmers in the city's formal fresh produce market (Mun Bbun and Thornton 2013). Specifically, researchers have found that, in greater 'Joburg', existing peri-urban land under cultivation by small-scale farmers, for example, in Roodepoort and Orange Farm, were being promoted by city officials as ideal sites for local urban food production, to be integrated into the City's fresh produce market initiative, which is part of the South African Government's wider National Fresh Produce scheme and expanding UA programme (Mun Bbun and Thornton 2013). South Africa's National Development Plan—2030 views peri-urban spaces as 'vibrant' locations for agricultural production of commodities such as flowers, vegetables, dairy and hydroponics (South African Government 2013: 9).

In general, urban farmers in South Africa are underappreciated at government and agricultural institutional levels (Malan 2015), and this must change by formally encouraging farmers' organisations and recognising contributions made by urban farmers to everyday urban life. Researchers argue that urban farmers can be more productive, through the provision of urban horticultural extension for technical training and development, identify and remove obstacles (e.g. bylaws) to UA and urban food marketing, and closing loops through the adoption of technology, for example, in converting food waste and waste water into valuable inputs (Thornton 2008; Richards and Taylor 2012; Malan 2015). The potential for UA to emerge as a policy tool for addressing structural inequalities, particularly in food policy space, is unlikely to occur through the efforts of urban social movements in South Africa. When compared to cities in other developing countries, urban activism has yet to emerge out of South Africa's spatially marginalised urban dwellers (Warshawsky 2014, 2016).

Urban Food Governance and 'Rights to the City'

Although these examples appear more 'top-down' than demonstrative of Lefebvre's call for urban dwellers to assert their 'rights to the city', they raise the following questions regarding the long-term implications of community–council partnerships, where alternative spaces become appropriated under the purview of city bylaws. This 'appropriation' may appear contradictory to core ideas in radical urban theory, the linking of local government—its resources and policymaking for structural change—with grassroots social production is perhaps a necessity to imagine a more sustainable urban future. In any case, UA is one form of social transformation taking place in 'North and South' cities. Government, or official, support for hitherto grassroots movements for socio-economic change can be beneficial through removing legislative barriers and providing infrastructure improvements and upgrades.

However, there is the hypothetical possibility that government involvement could be viewed as 'interfering' with actually existing alternative social spaces, which prefer to operate independently from bureaucracies attempting to regulate 'radical' activities. It could be that socially re-produced spaces could re-emerge elsewhere in the city, in reaction to undesired bureaucratic influences seeking to normalise hitherto 'alternative' responses to the global food system. This leads to an interesting question of what happens to alternative social movements when they become 'mainstream'? As Crane et al. (2013) suggest, these are questions ideal for critical urban geographers and radical social scientists, to take up the notion of collective urban rights and their potential power. Analyses of UA through the lens of Lefebvrian spatial production can advance critical understandings of alternative spatial production and the implications of its mainstream appropriation by the city (or state) for purposes of reinventing the city.

Urban Agriculture and the Globalised Food System

Interest in human and natural ecosystems as vital for urban sustainability are partly influenced by rapid growth in community-based 'urban greening' movements concerned with social, economic, environmental and food justice. For urban dwellers in low-income areas, globally, UA appears as a defiant act and symbolic of the failure of urban industrialism and the globalized food system. For urban food social movements, city planning has largely suppressed the development of urban agricultural activities, which is symptomatic of the tensions between actually existing neoliberalism and socially lived space. When alternative social spaces are integrated into mainstream urban planning as community development initiatives, what then happens to the meaning and purpose of the 'alternative movement'?

In zoning for UA, are rights to the city acknowledged and imagined urban geographies accepted and acted upon thus transforming or reordering urban spaces and reproducing social relations? Or do these relations become normalised, thus control and power in the use of urban food space continues as actual existing neoliberalism? Reconciling tensions in the contradiction between neoliberalism and alternatives might be possible through close community-driven consultation with their local councils.

The level of concern in civil society about the global food system, industrial or mechanised agriculture and its role in global warming have been raised in the popular consciousness, in books (Eric Schlosser's 'Fast Food Nation', Houghton Mifflin 2002) and documentaries (Robert Kenner's 'Food Inc' 2008) that stress the highly politicized, complex and interconnected processes involved in the production, marketing and distribution of food and its commodity chain. These issues are driving social movements with a 'green agenda' that seek solutions to minimise the impact of western consumption as a driver of climate change and its impact on communities in developing countries where much of the raw materials for western production and consumption are sourced. In some western communities, this socially driven 'green imperative' has developed into national 'Green Party' political groups. However, tension can emerge when social movements feel that their defining issues have been co-opted, or appropriated, by politics (seeing a 'political opportunity') and compromises the spirit and expectations of the collective will (Dryzek et al. 2003).

Applied research and institutional involvement in UA issues are now discussed alongside global concerns over climate change, 'peak oil' and related potential of a collapse in the global food system (Rosin et al. 2012; Thornton 2011). As a response to these challenges, UA is advanced as a strategy for improving, firstly, natural ecosystem services, such as carbon sequestration, through maintaining healthy soils and forests and, secondly, in 'human ecosystems' through 'urban greening' for healthy cities in moving towards closing the loop on high consumption of inputs (fossil fuels and food) and production of waste (Lehmann 2010).

Where economic outcomes appeal to both local producers and council mandates for local economic development, creating local food systems could help to break the impasse. Frameworks for local food systems, which both support local growers and marketing of produce, benefit low-income communities and make economically viable use of council, state and public lands, could provide pathways to compromises that are socially, environmentally and economically equitable.

The next chapter (Chapter 3), will present research on North and South UA, from my own previous work, as well as others. These studies were selected on the basis of their direct relevance to the themes specified throughout this book, namely, urban (food) governance, food security and the use of urban space for UA activities. What frameworks for city–community engagement exist in cities that have, seemingly, adapted forms of UA to formal urban policy and planning, and to what extent are they replicable to other cities (either within or external to a certain geographical and cultural context)?

REFERENCES

Acton, L. 2011. Allotment gardens: A reflection of history, heritage, community and self. *Papers from the Institute of Archaeology* 21: 46.

Adie, K. 2013. *Fighting on the homefront: The legacy of women in World War One.* London: Hodder & Stoughton.

Alkon, A.H. 2013. Food justice, food sovereignty and the challenge of neoliberalism. In *Food sovereignty: A critical dialogue*, September, 13–14.

Armar-Klemesu, M., and D. Maxwell. 2000. Accra: Urban agriculture as an asset strategy, supplementing income and diets. *Growing cities, growing food: Urban agriculture on the policy agenda*, 203–208.

Badami, M.G., and N. Ramankutty. 2015. Urban agriculture and food security: A critique based on an assessment of urban land constraints. *Global Food Security* 4: 8–15.

Barnett, C., P. Cloke, N. Clarke, and A. Malpass. 2005. Consuming ethics: Articulating the subjects and spaces of ethical consumption. *Antipode* 37: 23–45.

Barthel, S., J. Parker, and H. Ernstson. 2015. Food and green space in cities: A resilience lens on gardens and urban environmental movements. *Urban Studies* 52 (7): 1321–1338.

Battersby, J. 2012. Beyond the food desert: Finding ways to speak about urban food security in South Africa. *Geografiska Annaler: Series B, Human Geography* 94 (2): 141–159.

Bohstedt, J. 2014. Food riots and the politics of provisions in world history. *IDS Working Papers* 2014 (444): 1–31.

Borel-Saladin, J. 2017. Where to draw the line: Data problems and other difficulties estimating urbanisation in Africa. *Development Southern Africa* 34 (2): 137–150.

Bourne, G. 1942. Feeding post-war Europe. *Nature* 149: 182–184.

Brenner, N. 2012. What is critical urban theory? In *Cities for people, not for profit: Critical urban theory and the right to the city*, ed. N. Brenner, P. Marcuse, and M. Mayer, 11–23. New York: Routledge.

Brown, A. 2013. The right to the city: From Paris 1968 to Rio 2010. *International Journal of Urban and Regional Research* 37 (3): 957–971.

Bruntland, G.H. 1987. Our common future: Report of the World Commission on Environment and Development. World Commission on Environment and Development.

Cobbinah, P., M. Erdiaw-Kwasie, and P. Amoateng. 2015. Africa's urbanisation: Implications for sustainable development. *Cities* 47: 62–72.

Colding, J., and S. Barthel. 2013. The potential of 'urban green commons' in the resilience building of cities. *Ecological Economics* 86: 156–166.

Companioni, N., Y. Hernandez, E. Paez, and C. Murphy. 2002. The growth of urban agriculture. In *Sustainable agriculture and resistance: Transforming food production in Cuba*, ed. F. Funes, L. Garcia, M. Bourque, and P. Rosset. Oakland, CA: Food First.

Conservation Law Foundation. 2012. Growing green: Measuring benefits, overcoming barriers, and nurturing opportunities for urban agriculture in Boston. http://clf.org/growing-green/. Accessed 3 Feb 2014.

Crane, A., L. Viswanathan, and G. Whitelaw. 2013. Sustainability through intervention: A case study of guerrilla gardening in Kingston, Ontario. *Local Environment: The International Journal of Justice and Sustainability* 18 (1): 71–90.

Crush, J., and B. Frayne. 2011. Urban food insecurity and the new international food security agenda. *Development Southern Africa* 28 (4): 527–544.

Crush, J., A. Hovorka, and D. Tevera. 2011. Food security in Southern African cities: The place of urban agriculture. *Progress in Development Studies* 11 (4): 285–305.

Desai, M. 2014. *The paradigm of international social development: Ideologies, development systems and policy approaches.* New York: Routledge.

Detroit Free Press. 2011. Bill would create right to farm act exemption for Detroit [online]. Available at http://www.freep.com/article/20111128/NEWS06/111280346/Bill-would-create-Right-Farm-Act-exemption-Detroit. Accessed 14 Mar 2014.

Donald, B., and A. Blay-Palmer. 2006. The urban creative-food economy: Producing food for the urban elite or social inclusion opportunity? *Environment and Planning A* 38: 1901–1920.

Dryzek, J.S., D. Downes, C. Hunold, D. Schlosberg, and H.K. Hernes. 2003. *Green states and social movements: Environmentalism in the United States, United Kingdom, Germany, and Norway.* Oxford: OUP.

DuPuis, E.M., and D. Goodman. 2005. Should we go "home" to eat?: Toward a reflexive politics of localism. *Journal of Rural Studies* 21 (3): 359–371.

Eizenberg, E. 2012. Actually existing commons: Three moments of space in community gardens in New York City. *Antipode* 44 (3): 764–782.

Elden, S. 2004. Between Marx and Heidegger: Politics, philosophy and Lefebvre's 'the production of space'. *Antipode* 36 (1): 86–105.

Ellis, F., and J. Sumberg. 1998. Food production, urban areas and policy responses. *World Development* 26 (2): 213–225.

Endres, A.B., and J.M. Endres. 2009. Homeland security planning: What victory gardens and Fidel Castro can teach us in preparing for food crises in the United States. *Food and Drug Law Journal* 64: 405.

Foo, K., D. Martin, C. Wool, and C. Polsky. 2014. Reprint of "The production of urban vacant land: Relational placemaking in Boston, MA neighborhoods". *Cities* 40: 175–182.

Food and Agricultural Organisation (FAO) of the United Nations. 2015. *The state of food insecurity in the world* (*SOFI*). Rome: Food and Agriculture Organization (FAO) of the United Nations.

Frayne, B., W. Pendleton, J. Crush, B. Acquah, J. Battersby-Lennard, E. Bras, et al. 2010. *The state of urban food insecurity in southern Africa.* Urban Food Security Series, 2. Kingston and Cape Town: Queen's University and AFSUN.

Gibson-Graham, J.K. 2008. Diverse economies: Performative practices for 'other worlds'. *Progress in Human Geography* 32 (5): 613–632.

Goodman, D. 2003. The quality 'turn' and alternative food practices: Reflections and agenda. *Journal of Rural Studies* 19: 1–7.

Goodman, D., M. DuPuis, and M. Goodman. 2011. *Alternative food networks: Knowledge, practice and politics.* London: Routledge.

Guthman, J. 2008. Neoliberalism and the making of food politics in California. *Geoforum* 39 (3): 1171–1183.

Harvey, D. 2008. The right to the city. *The New Left Review* 53 (Sept–Oct): 23–40.

Hayden-Smith, R. 2014. *Sowing the seeds of victory: American gardening programs of World War 1.* Jefferson, NC: McFarland.

Henderson, B.R., and K. Hartsfield. 2009. Is getting into the community garden business a good way to engage citizens in local government? *National Civic Review* 98 (4): 12–17.

Hopkins, R. 2008. *The transition handbook: From oil dependency to local resilience.* Chelsea: Green Publishing.

Hossain, N. 2009. *Towards a new social justice agenda: Understanding political responses to crises.* IDS InFocus Policy Briefing, issue 11 October 2009. Brighton: Institute for Development Studies.

Huchzermeyer, M. 2012. Informal settlements: Production and intervention in Brazil and South Africa. *Latin American Perspectives* 29 (1): 83–105. New York: Sage.

Huchzermeyer, M. 2014. Humanism, creativity and rights: Invoking Henri Lefebvre's right to the city in the tension presented by informal settlements in South Africa today. *Transformation: Critical Perspectives on Southern Africa* 85 (1): 64–89.

IAASTD International Assessment of Agricultural Science, Technology and Development. 2009. *Agriculture at a crossroads: Global report.* Washington, DC: Island Press.

Islam, M., and N.C. Nag. 2010. *Economic integration in South Asia: Issues and pathways.* Chennai: Pearson Education.

Jarosz, L. 2009. The political economy of global governance and the world food crisis: The case of the FAO. *Review* 332 (1): 37–60.

Johannesburg Development Agency. 2015. Rooftop garden plants seeds of success. Available at: http://www.jda.org.za/index.php/corporate-social-responsibility/1542-rooftop-garden-plants-seeds-of-success2. Accessed 20 Oct 2017.

Kaufman, M., and J. Read. 2016. A case study: Legalizing commercial agriculture in Boston—A logical step towards integrating farming into urban life. In *Sowing seeds in the city*, ed. S. Brown, K. McIvor, and E. Hodges Snyder. Dordrecht: Springer.

Kneafsey, M. 2010. The region in food—Important or irrelevant? *Cambridge Journal of Regions, Economy and Society* 3 (2): 177–190.

Krishnan, S., D. Nandwani, G. Smith, and V. Kankarta. 2016. Sustainable urban agriculture: A growing solution to urban food deserts. In *Organic farming for sustainable agriculture*, 325–340. Cham: Springer.

Larder, N., K. Lyons, and G. Woolcock. 2012. Enacting food sovereignty: Values and meanings in the act of domestic food production in urban Australia. *Local Environment: The International Journal of Justice and Sustainability* 19 (1): 56–76. https://doi.org//10.1080/13549839.2012.716409.

Larder, N., K. Lyons, and G. Woolcock. 2014. Enacting food sovereignty: Values and meanings in the act of domestic food production in urban Australia. *Local Environment* 19(1): 56–76.

Lawson, L.J. 2005. *City bountiful. A century of community gardening in America.* Berkeley, Los Angeles, CA, and London, UK: University of California Press.

Lefebvre, H. 1991. *The production of space*. Malden, MA: Blackwell.

Lefebvre, H. 1996 [1968]. *Writings on cities*. Oxford: Blackwell.

Lehmann, S. 2010. *Principles of green urbanism: Transforming the city for sustainability*. London: Earthscan.

Lever, J., and A. Evans. 2017. Corporate social responsibility and farm animal welfare: Towards sustainable development in the food industry? In *Stages of corporate social responsibility*, 205–222. Cham: Springer.

Leyshon, A., R. Lee, and C. Williams (eds.). 2003. *Alternative economic spaces*. London: Sage.

Lourenco-Lindell, I. 1997. Food for the poor, food for the city: The role of urban agriculture in Bissau. *Geographical Journal of Zimbabwe* (28): 39–48.

Low, S.A., and S. Vogel. 2011. *Direct and intermediated marketing of local foods*. United States Department of Agriculture, Economic Research Report No. 128. U.S. Department of Agriculture, Economic Research Service.

Lynch, K., R. Maconachie, T. Binns, P. Tengbe, and K.S. Bangura. 2012. Meeting the urban challenge? Urban agriculture and food security in post-conflict Freetown, Sierra Leone. *Applied Geography* 36 (1): 31–39.

Lyson, T.A. 2004. *Civic agriculture: Reconnecting farm, food and community*. Metford: Tufts University Press.

Malan, N. 2015. Urban farmers and urban agriculture in Johannesburg: Responding to the food resilience strategy. *Agrekon* 54 (2): 51–75.

Marsden, T., J. Banks, and G. Bristow. 2000. Food supply chain approaches: Exploring their role in rural development. *Sociologia Ruralis* 40: 424–438.

Martellozzo, F., J.S. Landry, D. Plouffe, V. Seufert, P. Rowhani, and N. Ramankutty. 2014. Urban agriculture: A global analysis of the space constraint to meet urban vegetable demand. *Environmental Research Letters* 9 (6): 1–9.

Martin, G. 2011. Going local: Quantifying the economic impacts of buying from locally owned businesses in Portland, Maine. Maine Center for Economic Policy.

Martinez, S., M. Hand, M. Da Pra, S. Pollack, K. Ralston, T. Smith, ... and C. Newman. 2010. *Local food systems: Concepts, impacts, and issues* (No. 24313), University Library of Munich, Germany.

Maxwell, D.G. 1994. The household logic of urban farming in Kampala. In *Cities feeding people: An examination of urban agriculture in East Africa*, ed. A.G. Egziabher, D. Lee-Smith, D.G. Maxwell, P.A. Memon, L.J.A. Mougeot, and C.J. Sawio. Ottawa: IDRC.

Maxwell, D., C. Levin, M. Armar-Klemesu, M. Ruel, and S. Morris. 2000. *Urban livelihoods and food and nutrition security in Greater Accra, Ghana*. Washington, DC, U: IFPRI.

Maye, D., and J. Kirwan. 2010. Alternative food networks. *Sociology of Agriculture and Food* 20: 383–389.

Maye, D., L. Holloway, and M. Kneafsey (eds.). 2007. *Alternative food geographies: Representation and practice*. Oxford: Elsevier.

Mbiba, B. 2000. Urban agriculture in Harare: Between suspicion and repression. In *Growing cities, growing food: Urban agriculture on the policy agenda—A reader on urban agriculture*, ed. N. Bakker, M. Dubbeling, S. Gündel, U. Sabel-Koschella, and H. de Zeeuw, 285–301. Feldafing: GTZ.

McKernan, M. 1983. All In! Australia during the second world war. *Nelson*.

Mougeot, L.J.A. 1994. *Urban food production: Evolution, official support and significance*. Cities Feeding People Report 8. Ottawa: IDRC.

Mougeot, L. 1999. For self-reliant cities: Urban food production in a globalizing south. In *For hunger-proof cities: Sustainable urban food systems*, ed. M. Koc et al. Ottawa: IDRC.

Mougeot, L. 2000. Urban agriculture: Definition, presence, potentials and risks. In *Growing cities, growing food: Urban agriculture on the policy agenda*, ed. N. Bakker, M. Dubbeling, S. Gündel, U. Sabel-Koschella, and H. de Zeeuw, 1–42. Zentralstelle fuer Ernaehrung und Landwirtschaft. Feldafing: German Foundation for International Development.

Mun Bbun, T., and A. Thornton. 2013. A level playing field? Improving market availability and access for small-scale producers in Johannesburg, South Africa. *Applied Geography* 36 (1): 40–48.

Ncube, N., and D. Ncube. 2016. *Urban agriculture and food security: A case study of old pumula suburb of Bulawayo in Zimbabwe*.

Nel, E., G. Hampwaye, A. Thornton, C. Rogerson, and L. Marais. 2009. Institutional responses to decentralization, urban poverty, food shortages and urban agriculture. GDN Working Paper Series, Working Paper No. 36, Global Development Network (GDN), 1–29.

Nugent, R. 2000. The impact of urban agriculture on the household and local economies. In *Growing cities, growing food: Urban agriculture on the policy agenda*, ed. N. Bakker et al. Feldafing: Deutsche Stiftung fuer internationale Entwicklung.

Oosterveer, P., and D.A. Sonnenfeld. 2012. *Food, globalization and sustainability*. London: Routledge.

Orsini, F., R. Kahane, R. Nono-Womdim, and G. Gianquinto. 2013. Urban Agriculture in the developing world: A review. *Agronomy for Sustainable Development* 33 (4): 695–720.

Pack, C.L. 1917. Urban and suburban food production. *Annals of the American Academy of Political and Social Science* 74: 203–206.

Pack, C.L. 1919. *The war garden victorious*. Philadelphia: J.B. Lippincott Company.

Popovitch, T. 2014. 10 American cities lead the way with urban agriculture ordinances. Available at http://seedstock.com/2014/05/27/10-american-cities-lead-the-way-with-urban-agriculture-ordinances/. Accessed 29 May 2014.

Poulsen, M.N., P.R. McNab, M.L. Clayton, and R.A. Neff. 2015. A systematic review of urban agriculture and food security impacts in low-income countries. *Food Policy* 55: 131–146.

Purcell, M. 2002. Excavating Lefebvre: The right to the city and its urban politics of the inhabitant. *GeoJournal* 58 (2–3): 99–108.

Purnomohadi, N. 2000. Jakarta: Urban agriculture as an alternative strategy to face the economic crisis. In *Growing cities, growing food: Urban agriculture on the policy agenda*, ed. N. Bakker, M. Dubbeling, S. Gündel, U. Sabel-Koshella, and H. de Zeeuw, 453–465. Feldafing, Germany: Zentralstelle für Ernährung und Landwirtschaft (ZEL).

Quick, S. 1977. Bureaucracy and rural socialism in Zambia. *The Journal of Modern African Studies* 15 (3): 379–400.

Rakodi, C. 1988. Urban agriculture: Research questions and Zambian evidence. *The Journal of Modern African Studies* 26 (3): 495–515.

Richards, R., and S. Taylor. 2012. Changing land use on the periphery: A case study of urban agriculture and food gardening in Orange Farm. The South African Research Chair in Spatial Analysis and City Planning. Johannesburg: The University of the Witwatersrand. http://hdl.handle.net/10539/17142.

Rogerson, C.M. 1993. Urban agriculture in South Africa: Policy issues from the international experience. *Development Southern Africa* 10 (1): 33–44.

Rogerson, C. 2003. Towards 'pro-poor' urban development in South Africa: The case of urban agriculture. *Acta Academica Supplementum* 1: 130–158.

Rosset, P., and D. Benjamin. 1994. *The greening of Cuba: A national experiment in organic agriculture.*

Rosin, C., P. Stock, and H. Campbell (eds.). 2012. *Food systems failure: The global food crisis and the future of agriculture.* London: Earthscan.

Sahn, D.E. 1989. A conceptual framework for examining the seasonal aspects of household food security. In *Seasonal variability in third world agriculture: The consequences for food security*, ed. D.E. Sahn. Baltimore: John Hopkins University Press.

Sanyal, B. 1987: Urban cultivation amidst modernisation: How should we interpret it? *Journal of Planning Education and Research* 6: 187–207.

Schlosser, E. 2002. *Fast food nation: The dark side of the All-American meal.* Houghton Mifflin.

Simon, D. 1995. Debt, democracy and development: Sub-Saharan Africa in the 1990s. In *Structurally adjusted Africa: Poverty, debt and basic needs*, ed. D. Simon, W. van Spengen, C. Dixon, and A. Närman. Pluto, London.

Smit, J., A. Ratta, and J. Nasr. 1996. *Urban agriculture: Food, jobs and sustainable cities.* New York: UNDP.

Smit, J., A. Ratta, and J. Nasr. 2001. *Urban agriculture: Food, jobs and sustainable cities*, 2001st ed. New York: UNDP.

Smith, D., and D. Tevera. 1997. Socio-economic context for the householder of urban agriculture in Harare, Zimbabwe. *Geographical Journal of Zimbabwe* 28: 25–38.

Smith, A., and J.B. McKinnon. 2007. *The 100-mile diet: A year of local eating.* Toronto: Random House Canada.

Sneyd, L., A. Legwegoh, and E. Fraser. 2013. Food riots: Media perspectives on the causes of food protest in Africa. *Food Security* 5 (4): 485–497.

South African Government. 2012. National Development Plan—2030, Ch 8–9. Available at: http://www.gov.za/sites/www.gov.za/files/devplan_ch8_0.pdf. Accessed 28 Sept 2017.

Tawodzera, G. 2011. Vulnerability in crisis: Urban household food insecurity in Epworth, Harare, Zimbabwe. *Food Security* 3 (4): 503–520.

Thaman, R. 1975. *Urban gardening in Papua, New Guinea and Fiji: Present status and implications for urban land-use planning.* Suva: University of the South Pacific.

Thornton, A. 2008. Beyond the metropolis: Small town case Studies of urban and peri-urban agriculture in South Africa. *Urban Forum* 19 (3): 243–262.

Thornton, A 2011. Food for thought? The potential of urban agriculture in local food production for food security in the South Pacific. In *Food systems failure: The global food crisis and the future of agriculture*, ed. C. Rosin, P. Stock, and H. Campbell, 203–218. London: Earthscan.

Thornton, A. 2012. *Urban agriculture in South Africa: A study of the Eastern Cape Province.* Lewiston, NY: Edwin Mellen Press.

Thornton, A. 2017. "The Lucky country"? A critical exploration of community gardens and city–community relations in Australian cities. *Local Environment* 22 (8): 969–985.

Thornton, A. 2018. Food security in African Cities. In *The Routledge Handbook of African development*, ed. T. Binns, K. Lynch, and E. Nel. London: Routledge.

Thornton, A., and C. Rogerson. 2013. African cities and the millennium development goals: A case for applied geography. *Applied Geography* 36 (1): 1–3.

Thornton, A., K. Lyons, and S. Sharpe. 2018. Community gardens for social and food justice: The case of urban agriculture in Australian cities. In *International handbook of community development*, ed. L. Shevellar and P. Westoby. London: Ashgate.

Thornton, A., J. Momoh, and P. Tengbe. 2012. Institutional capacity building for urban agriculture research using Participatory GIS in a post-conflict context: A case study of Sierra Leone. *Australasian Review of African Studies* 33 (1): 165–176.

Thornton, A., E. Nel, and G. Hampwaye. 2010. Cultivating Kaunda's plan for self-sufficiency: Is urban agriculture finally beginning to receive support in Zambia? *Development Southern Africa* 27 (4): 613–625.

UN-Habitat. 2015. Governance. Available at: http://unhabitat.org/urban-themes/governance/. Accessed 20 Oct 2017.

United Nations Development Programme (UNDP). (2015a). The post-2015 development agenda. Available at http://www.undp.org/content/undp/en/

home/mdgoverview/post-2015-development-agenda.html. Accessed 05 Dec 2017.

UNDP. (2015b). SDG Goal 11: Cities will play an important role in achieving the SDGs. Available at http://www.my.undp.org/content/malaysia/en/home/presscenter/articles/2015/08/04/sdg-goal-11-cities-will-play-an-important-role-in-achieving-the-sdgs.html. Accessed 05 Dec 2017.

Van Veenhuizen, R., G. Prain, and H. De Zeeuw. 2001. Appropriate methods for urban agriculture research, planning implementation and evaluation. *Urban Agriculture Magazine* 5: 1–5.

Venn, L., M. Kneafsey, L. Holloway, R. Cox, E. Dowler, and H. Tuomainen. 2006. Researching European 'alternative' food networks: Some methodological considerations. *Area* 38 (3): 248–258.

Warner, S. 1987. *To dwell is to garden: A history of Boston's community gardens.* Boston: Northeastern University Press.

Warshawsky, D. 2014. Civil society and urban food insecurity: Analyzing the roles of local food organizations in Johannesburg. *Urban Geography* 35 (1): 109–132.

Warshawsky, D. 2016. Civil society and the governance of urban food systems in sub-Saharan Africa. *Geography Compass* 10 (7): 293–306.

Weiss, E. 2008. *Fruits of victory: The Woman's Land Army of America in the Great War.* Washington, DC: Potomac Press.

Wenban-Smith, H., A. Faße, and U. Grote. 2016. Food security in Tanzania: The challenge of rapid urbanisation. *Food Security* 8 (5): 973–984.

World Resources Institute. 2000. *A guide to world resources 2000–2001, people and ecosystems: The fraying web of life.* Washington DC: World Resources Institute.

Yi-Zhang, C., and Z. Zhangen. 2000. Shanghai: Trends towards specialised and capital-intensive urban agriculture. *Growing cities growing food: Urban agriculture on the policy agenda*, 467–477. Feldafing: DSE-ZEL.

Zezza, A., and L. Tasciotti. 2010. Urban agriculture, poverty, and food security: Empirical evidence from a sample of developing countries. *Food Policy* 35 (4): 265–273.

CHAPTER 3

City Case Studies in Urban Governance and Urban Activism

Abstract This chapter will present UA case studies in the global North and South, from my own research (previous and current), as well as others. These studies were selected on the basis of their direct relevance to the themes specified throughout this book, namely, urban governance, food security and the use of urban space for UA activities. This chapter will begin with an overview of UA practice in cities in developing and developed countries. It will then address the following questions, and apply these to the case studies: (1) are there existing frameworks or formal processes for city–community engagement in urban food space issues and (2) to what extent are they replicable to other cities (with due consideration for how geography, culture and history influence places and spaces)?

Keywords Community · Governance · Activism · Urban
Social gardens · Berlin · Boston · Canberra · Sydney
Urbanisation · Urban space

Urban Agriculture Practice in the Global South & Economies in Transition: An Overview of the Americas, Asia, Caribbean and Eastern Europe

Urban agriculture in the developed world appears to have embraced a more holistic and socially conscious path to urban planning, with positive social, economic and environmental impacts. In the developing world,

© The Author(s) 2018
A. Thornton, *Space and Food in the City*,
https://doi.org/10.1007/978-3-319-89324-2_3

it appears that urban agriculture emerged primarily as a local level or grass roots response to the inability of governments, the economy and society to ensure consistent, low-cost food supply in a climate of economic and social instability. Consequentially, state authorities in developing countries increasingly view sustainable urban development as the key to political, economic and social stability (Hamilton et al. 2014). In the following paragraphs, a discussion of how urban agriculture is practiced in developing countries highlights internal local level, bottom-up (development from below) poverty responses and external, or institutional, top-down (development from above) approaches to urban challenges and sustainability.

Urbanisation and Urban Sprawl

In terms of addressing the challenges of urbanisation, urban agriculture is cited to have made a positive impact through the cultivation of peri-urban land as a means to control urban sprawl and preserve land resources in several countries. Earlier studies of urban agriculture in Mexico City, Mexico (Torres-Lima et al. 2000: 374), found that city officials tolerated the practice, although it was not legally permitted, as it was seen to complement and engage with the informal and formal economy. Further, government agencies and commissions had acknowledged its importance for the preservation of the environment and food security through the establishment of 'green-belts', consisting of an ecological park and the promotion of organic 'backyard' gardening (Torres-Lima et al. 2000: 382). More recent studies (Dieleman 2017) of urban agriculture in Mexico City suggest that it may have found its niche, through its formal integration into environmental policy, where it may contribute to the ecological infrastructure of the city.

In China (Yi-Zhang and Zhangen 2000: 467), the city of Shanghai is nearly self-sufficient in food, mainly from large peri-urban gardens. As a response to urbanisation, 80% of arable land is protected under the *Agricultural Protection Law of 1998*. Moreover, the Chinese Government implemented UA-friendly policies over 20 years ago that encourages household food production (Fan et al. 2004; Hamilton et al. 2014). Peri-urban market gardening is viewed as 'indispensable' for urban food security, but rapid urban sprawl is a threat to productive peri-urban lands (Hamilton et al. 2014: 60). The government pursues capital and labour-intensive, mechanised (for high output) and high input agricultural development as a means to maintain social stability, green spaces and reduce air pollution (ibid.). In addition, UA is marketed as a form of agri-tourism, Yang et al. (2010, from Hamilton et al. 2014: 60), give the example of Xiedao Green Resort in Beijing, where "90% of its area is devoted to agricultural production and 10% to tourism, including accommodation".

In contrast, authorities in Hubli-Dharwad, India do not officially recognise, support or encourage urban agriculture. However, the Hubli-Dharwad Urban Development Authority does maintain a 'green-belt' around the city where agricultural land remains, but developers can request zoning changes in areas considered within the 'green-belt' (Nunan 2000: 449). Nonetheless, as in many developing countries, urban and peri-urban agriculture is seen as a survival strategy for the urban poor, as well as one way to earn an income. In Delhi, much of the urban food production takes place on the Yamuna River floodplain, while an additional 44% is carried out in the peri-urban areas and on rooftops (Hamilton et al. 2014).

Somewhat mirroring the land-tenure situation in post-apartheid South Africa (land ownership rights were primarily restricted to whites under apartheid—arable land in particular) is the process of land restitution in Bulgaria. Following the collapse of communism land restitution has been slow and difficult (where the state had previously assumed possession of all privately owned land). Most of the problems involved transfer of land near large cities, which became "villa zones" (Yoveva et al. 2000: 513). For example, a typical scenario being that a pre-socialist owner and the 'new', post-socialist owner engage in a dispute over buildings that have since been built on a given plot, which both parties claim ownership. Either the new owner must buy the land back, or the previous owner must buy the building from the new owner. Where 42.4% of the land has not been returned, most farming is done with temporary land use rights, which has resulted in a non-existent market for agricultural land, inhibiting long-term investments in improving the land and making it difficult to obtain credit for agriculture, as cultivated land is not accepted as collateral (Yoveva et al. 2000: 513). As a result, the practice of urban agriculture, although popular among all income groups, has experienced a decline (ibid.).

Economic Crises

Food insecurity brought on by economic crises has often necessitated a governmental response to support, facilitate and ensure local production of food and, in some cases, self-sufficiency in urban food supply.

In Havana, Cuba (Companioni et al. 2002; Hamilton et al. 2014; Hallett et al. 2016) and Sofia, Bulgaria (Yoveva et al. 2000), official recognition, support and governmental initiatives for urban agriculture have absorbed the economic shocks that followed the collapse of the Soviet Union and, consequently, the Eastern Bloc economic community (COMECON). Further, in the case of Cuban and Bulgarian cities, strict hygiene and sanitary standards regulate and monitor the marketing of urban agriculture products in urban open markets and auctions, which include on-site epidemiological laboratories (Yoveva et al. 2000: 512) and phyto-sanitary services (Novo and Murphy 2000: 340).

In Lima, Peru (Dasso and Pinzas 2000; Prain et al. 2010), Cagayan de Oro and Mati City, Philippines (Potutan et al. 2000; Hamilton et al. 2014[1]), and to some extent Jakarta, Indonesia (Purnomohadi 2000; Hamilton et al. 2014), some municipalities are supportive of types of urban agriculture and coordinate activities with local communities and non-governmental organisations. For example, complementary to the role of municipal government, is the important role of local television and the print media in creating awareness and promoting support for urban agriculture in Cagayan de Oro (Potutan et al. 2000: 423).

Household Response to Poverty

Although variations exist, household responses to poverty in the following examples reveal complex social arrangements and integrated methods of production. Rural migrants to the Bolivian city of La Paz (Kreinecker 2000: 392) remain connected to a system of Andean reciprocity where goods and mutual support are exchanged in a socially complex rural–urban network (ibid.). The exchange of labour and goods assists rural migrants to adapt their rural agricultural skills to an urban environment. Urban agriculture practitioners in Hubli-Dharwad, India also maintain their traditional and rural links by raising livestock in the city (Nunan 2000: 429). Urban dwellers may invest in mango orchards in the rural areas to alleviate tax burden or bring in fodder for livestock (Nunan 2000: 441). This type of urban agriculture activity, in many cases, has been "passed on from one generation to the next" (Nunan 2000: 435). Urban agriculture, as a result, "exists in every corner, near the centre,

---

[1]City Farmer. 2017. http://www.cityfarmer.info/2017/05/07/philippines-mati-city-promotes-backyard-gardening/.

as well as on the outskirts of the city" (Kreinecker 2000: 395); and "vegetables are grown on railway tracks and in small patches of open land" (Nunan 2000: 430).

The informal economy in La Paz accounts for 70% of the national economy, of which urban agriculture figures prominently, with an estimated 30% of La Paz's agricultural requirements being provided by urban agriculture (Nunan 2000: 395). Small animals are raised and fruits and vegetables are grown primarily for subsistence reasons in peri-urban and backyard gardens in Sofia, Bulgaria (Yoveva et al. 2000: 507). Yoveva et al. (2000) claim that urban agriculture in Bulgaria is centuries old; although, in the transition from communism to capitalism, the newly rich no longer cultivate from home but seek "clean food products from the farmers in the neighbourhood" (Yoveva et al. 2000: 513). Many Asian observations of urban agriculture processes point out the contribution of backyard gardens to meeting the food needs of practicing households. These observations include Hanoi, Vietnam, where 80% of fresh vegetables, 50% of meat and 40% of eggs originate from urban and peri-urban areas (De Haen 2002). Similarly, in Shanghai, China, 60% of vegetables, 50% of meat and nearly 90% of milk and eggs are provided in urban and peri-urban areas (ibid.).

Moreover, most large Chinese cities, such as Hong Kong, are nearly self-sufficient in perishable food crops and the Shanghai municipal government has a fully integrated [metropolitan-region] food supply system (Mougeot 1994). Overall, Asia is viewed as the most well-developed urban agriculture region, "In the late 20th century, the greatest advances in production and marketing systems for urban agriculture are found in and around major Asian cities," where policymakers and planners, for decades, have been overtly promoting it as a critical urban function (Mougeot 1994: 2).

CHARACTERISTICS OF URBAN AGRICULTURE PRACTITIONERS IN AFRICA

Significance and Limitations of Urban and Peri-Urban Agriculture

In reviewing earlier literatures on urban agriculture in the African context, its emergence is often discussed as a response to economic crises, which, in many cases, resulted from the implementation of structural adjustment programmes (Maxwell 1994; Gertel and Samir 2000).

For the most part, urban agriculture is practiced by all income groups and is of crucial importance to the poorest households for subsistence (Thornton 2012). This role is recognised, for example, in Dar es Salaam, where urban agriculture has been discussed as part of urban planning (Foeken et al. 2004). Currently urban farmers must 'fend for themselves', as UA remains informally regulated in many urban wards (Mkwela 2014: 45). Given the region's harsh urban economic conditions, urban agriculture has long been seen as an "economic necessity" for the poorest households and the average middle-income earner views urban agriculture as the "logical thing to do" (Sawio 1994: 23). However, the potential impact of urban agriculture on food security, particularly in terms of amount of urban land availability, in many African countries is argued as being quite limited (Badami and Ramankutty 2015). As such, a more likely scenario for UA as a livelihood strategy is largely supplemental, for food or cash, among the poorest households.

Food production for household food security in Kampala, Uganda is the most common type of urban agriculture. However, "the food produced does not constitute the majority of what a household consumes", as the majority of what a household consumes comes from the market, but the garden provides a buffer against shortages in food or cash (Maxwell 1994: 55). The 2007–2011 food crisis period, emphasised the importance of UA to food security in Ugandan cities. For example, at the level of local government the "Kampala Capital City Authority (KCCA) now recognizes UA as a land use system and a vital policy issue"; however, the policy environment is still evolving (Sabiiti and Katongole 2014: 238).

For urban residents in Cairo, Egypt (Gertel and Samir 2000: 214), rural areas provide urban markets with comparatively low-prices throughout the year. A typical crop cultivated in urban areas, clover, yields high prices as fodder for urban livestock. The preference for livestock rearing in urban areas is linked to local concerns over increases in food price and preferences for urban animal husbandry in economic development (Ibrahim and Elariane 2018).

In a study of urban agriculture in Harare, Zimbabwe, Mbiba (2000: 291) determined that urban agriculture features as a "tiny component" of a diverse range of informal survival activities for the urban poor (Mbiba 2000: 291). However, approximately one year later, Mbiba (2001: 19) noted that, as a result of "poor planning institutions", urban agriculture in Harare, Zimbabwe has "taken over" from Lusaka, Zambia as "the capital city of urban agriculture in Africa" (assigned earlier to Lusaka in a study by Sanyal 1987). Gogwana (2001: 58) also finds that urban agriculture is an "important socio-economic activity, particularly for the poor". Previous studies had deemed UA as an insignificant activity for even the poorest households in Zimbabwe, though it is practised by various income groups (Smith and Tevera 1997; Mbiba 2000). Masvaure (2016: 207) argues that these earlier studies of UA in Zimbabwe focussed largely on field or off-plot farming, and did not include UA activities taking place at the household (on-plot). As a result, much of the activity was overlooked. Masvaure (2016: 209) finds that UA is more important today, than it was in the past, "Farmers who started farming in the 1970s and 1980s attested to the fact that urban agriculture was not as crucial then as it is today", which is attributed to economic decline.

Rural–Urban Migration

Some observers claim that urban agriculture practitioners are recent migrants from rural areas who depend on their agricultural or traditional skills to survive in the urban areas (Mayer 1971; Bundy 1979; Potter and Unwin 1989; Tacoli 1998; Gogwana 2001). Conversely, other observers argue that African urban farmers are rarely recent migrants (Freeman 1991; Egziabher 1994; Sawio 1994; Mbaye and Moustier 2000; Jacobi et al. 2000). In Dar es Salaam, a majority of urban farmers have been living in town for at least 10–15 years (Sawio 1994; Jacobi et al. 2000; Foeken et al. 2004). Still, many urban agriculture practitioners in Ghana are recent male migrants from the north of Ghana (Armar-Klemesu and Maxwell 2000: 197), while in Nairobi, Kenya, urban agriculturalists remain close to their rural roots, cultivating in both urban areas where they live and in the rural areas (Lee-Smith and Ali Memon 1994).

Age of Practitioners

Earlier studies have established the claim that urban agriculture has been increasing among all age groups in response to decreasing employment opportunities and purchasing power (Egziabher 1994; Maxwell 1994; Sawio 1994; Mlozi 1996; Mtani 1997; Gertel and Samir 2000; Mbaye and Moustier 2000; Jacobi et al. 2000). For elderly or aged urban dwellers, urban agriculture allows pensioners to stretch their household budget by growing some of what they eat (Thornton 2008). The role of urban and peri-urban agriculture can increase when pensioners take responsibility for grandchildren orphaned by HIV/AIDS (Mubvami et al. 2006; Hungwe 2007; Mujere 2017).

Several observers claim that urban agriculture has been increasing among all age groups (although retirees rarely cultivate) in response to decreasing employment opportunities and purchasing power (Egziabher 1994; Maxwell 1994; Sawio 1994; Mlozi 1996; Mtani 1997; Gertel and Samir 2000; Mbaye and Moustier 2000; Jacobi et al. 2000).

The Role of Women

Overall, with the exception of Accra, Ghana (Armar-Klemesu and Maxwell 2000), women are more likely to engage in urban agriculture, both at home and in community gardens, to supplement the household food supply and prevent child malnutrition (Sawio 1994; Mlozi 1996; Mtani 1997; Gertel and Samir 2000; Maxwell et al. 2000; Mbiba 1995, 2000). The prevalence of urban agriculture practitioners in Kampala, Uganda was found evenly distributed among men and women (Maxwell 1994). Ethiopian women, due to the traditional system of household membership and headship, are largely responsible for the needs of the household (Egziabher 1994). In a study of households involved with communal or cooperative gardens, women (mothers and daughters) would tend to the private household garden, while men work in the communal or cooperative garden. Single mothers were found to have a double burden, where they would work at the private plot and in the cooperatives (ibid.). In contrast, low-income single women with children in Nairobi, Kenya only cultivate as a last resort to finding employment. With limited opportunities for employment, 56% of urban agricultural-ists in Kenya are women; in the capital city of Nairobi, it is 62% (Lee-Smith and Ali Memon 1994), a fact attributed to the relatively low-level of education in comparison with men.

Production Systems

Inner-city home vegetable production for home consumption is the most common production system in Ghana (Armar-Klemesu and Maxwell 2000: 187), Tanzania (Mlozi 1996; Mtani 1997; Jacobi et al. 2000; Foeken et al. 2004), Ethiopia (Egziabher 1994), Kenya (Lee-Smith and

Ali Memon 1994) and Uganda (Maxwell 1994). Land tenure-security problems are also common but do not represent an impassable barrier to urban agriculture (Mbaye and Moustier 2000; Jacobi et al. 2000). Intra-urban open spaces and peri-urban (former rural farmland) areas are commonly leased to cultivators by landowners, with the produce geared towards the local market in places such as Dakar, Senegal (Mbaye and Moustier 2000) and Dar es Salaam, Tanzania (Jacobi et al. 2000). The raising of poultry is "by far the most important" urban agriculture activity in Cairo, almost exclusively undertaken by low-income groups and women (Gertel and Samir 2000: 217–218). Due to space constraints, most Cairo residents raising poultry prefer to live on the top floors of buildings (70.8%), in order to have access to rooftops (ibid.). Urban agriculture appears to be a small-scale subsistence activity, requiring limited inputs (mainly manure for fertiliser) and labour. Therefore, the role of commodity exchanges is largely confined to "a simple self-sufficient peasant economy" in small towns and petty informal commodity exchanges in larger towns and cities (Lee-Smith and Ali Memon 1994: websource; also in Maxwell 1994). In Ethiopian households, traditional extended family systems create a form of household "self-insurance" for labour and assistance when needs arise (Egziabher 1994: websource).

Overall, urban agriculture in global South cities appears to be significant for a wide socio-economic range of households, using a variety of production systems. The types of production systems used seem to depend on the plot location and size and are oriented towards the needs of the practitioner rather than the market. Based on the examples above, urban agriculture appears to offer some degree of relief to households (in terms of sustenance and subsidising income spent on food) experiencing extreme poverty. These activities also exist in absence of government service provision, outside of the market economy and are symbolic of embedded urban poverty in the developing world. In the context of urban governance and social justice, urban agriculture differs by comparison in North and South cities. In the former, UA and its various forms are championed by urban activists for social and spatial justice, where the social production of (urban food) space is an act of resisting 'actually existing' neoliberalism in the spaces of everyday life. In the latter, it is an embattled concept and survivalist struggle against 'actually existing' desperation in the spaces of everyday life.

OVERVIEW UA CASE STUDIES FROM GLOBAL NORTH CITIES

Australia[2]

Australia is one of the world's most urbanised countries, with 90% of the population currently residing in cities and towns. Sydney and Melbourne have populations over 4 million; and by 2030 this figure may reach 6 million. The ability of Australia to feed its share of the global population is already challenged, and will become even more so given the population growth trends. Australia already imports much of its food supply (while exporting 77% of its food crops[3]), and this is vulnerable to global shocks in production and distribution.

Australia's agricultural space in the built-up and peripheral 'fringes' are being lost to housing developments, in particular to those that do not plan for open green space. Despite this, urban agriculture is getting a closer look by city planners, particularly in Melbourne and Sydney, as a form of local food production. In Melbourne's suburbs, city officials have approved street-side planter boxes and several city councils are employing officers to facilitate the application of urban agriculture (Melbourne Planning Authority 2008). In Sydney, the Environment and Heritage Committee is funding an investigation into potential sites and models for a 'Sydney City Farm', sited on council land, and Sustainable Living Centre (Australian City Farms and Community Gardens Network 2011). Similarly, Canberra City Farm, part of the Urban Agriculture Australia Network, had successfully negotiated a 10-year lease with the ACT Government (Australian Capital Territory, ACT) to establish a city farm and sustainability education centre on a nature reserve in Canberra, the nation's capital city.

In terms of the spatial and physical capacity of Australian cities and peripheries to produce enough vegetables in relation to real consumption patterns, Australia is one of 22 countries (from a sample size of 165 cities) that would require less than 10% of their cultivable urban area to satisfy urban vegetable demand, due to low urban population densities (Martellozzo et al. 2014). This can be placed in the context of cities

---

[2] This Section Draws from My Own Previously Published: Thornton, A. 2017. "The Lucky country"? A critical exploration of community gardens and city–community relations in Australian cities. *Local Environment* 22 (8): 969–985.

[3] National Farmer Federation. 2017. http://www.nff.org.au/farm-facts.html.

with higher urban densities in US cities (discussed below), where extensive zoning for urban agriculture has been prioritised by city officials
and supported by residents to expand local food production. Compared
to many North American cities, Australian cities have some of the lowest population and residential development densities (Gurran 2011). It
would appear that Australian cities have untapped potential to develop
local food networks through encouraging urban agriculture in inner city
and peripheral spaces.

However, partly due to 'fast-tracking' of development propos-
als, city planners in Australia appear slow to advance notions of urban
environmental performance with 'liveability' and deriving various social
benefits from its open spaces (Gaynor 2006; Gurran 2011; Freestone
2014). Despite this, at the grassroots level, several groups are proac-
tively advocating the social use of urban space for local food production
and marketing, as forms of community pride and urban environmentalism (Australian Food Sovereignty Alliance, Australian City Farms and
Community Gardens Network, Urban Agriculture Australia, to name a
few). These groups, and others, argue that supporting local food production and marketing networks can have multiplier effects in stimulating
local sustainable economic development, including job creation.

In Thornton's (2017) study of city–community relationships and community food spaces, councils are hesitant to identify and remove existing
barriers to community urban food space. Moreover, shifting or unrelia-
ble community gardens policies are not reconciling market-orientated
urban space with the desired social use of space at the grassroots level.
Thornton's findings also indicate Australian communities share charac-
teristics that reflect broader understandings of Gibson-Graham's (2008)
diverse economy (discussed above), with respect to alternative, non-market and non-capitalist transactions, where people volunteer or participate
in alternative food networks (or other than supermarkets) and volunteer
time and assets to food gardens (cultivation, 'working bees' and community food education, nutrition awareness) and farmer's markets.

Community gardeners and grassroots civil society organisations in
Australian cities are increasingly calling for access to public, council or
state land to create and facilitate resilient, equitable and diverse local
food systems. Thornton (2017) suggests that city officials have not
gone far enough, to engage in cooperative relationships with communities seeking alternative and diverse urban food systems. This is indicative
of a Lefebvrian criticism of city regulators for creating an urban policy

environment that is detrimental to 'rights to the city' claims (Lefebvre 1991). This critique appears quite relevant to Australian cities, where urban land use decisions appear fixated on serving neoliberal interests, at the expense of innovative city–community partnerships for healthy 'liveable' cities (Australian Government 2013). The following interviews from Thornton (2017: 976–979) with community garden activists highlight this critique.

In contrast to the Berlin case study (below), a culture of urban food gardening is weak. This is not to say that urban food activists and advocacy groups in Berlin, and in other German cities, have unwavering support for their urban food activities. In an era of rapid urbanisation and escalating value of inner-city properties and financially constrained municipalities, there is a struggle to maintain an urban food growing culture in European and North American cities (dating back to the kleingärtens, or Schrebergärtens, and 'war' and 'victory gardens') and the spaces to practice it. However, where these activities are more prominent in European and many North American cities, existing frameworks or formal processes for city–community engagement in urban food space issues do exist. This is not the case in Australia, where a more ad hoc approach is taken by city councils to request for space from community food gardening enthusiasts. In Thornton (2017), members of community gardens in Sydney and Canberra emphasised difficulties in working with local councils/municipalities as the main barrier to starting up or maintaining urban gardening activities. Council staffers were often described as having interests and priorities that were incongruent with the interests of garden advocates. In Sydney, for example, the historic Le Perouse Market Garden (initiated a century ago by Chinese immigrants) was described as being under threat of cemetery expansion driven by the Greek Orthodox community. This situation was explained by a campaigner engaged in activism to save the Market Garden (Thornton 2017: 978):

> There was a councillor [who argued that] food gardens are insignificant, [as] they don't contribute to food security; it doesn't mean anything. [They miss the point that] of course 7 acres is not going to contribute [to food security]. [The idea is that] one 7 ha garden can lead to another 7 ha, and then another 7 ha and so on. The number of people is what matters, to contribute to reducing emissions of food transport, internationally and domestically.

She further explained:

> If I had the time and resources, I could bring together more like-minded
> people to run a smart campaign with key people. Trying to rally support
> for market gardens with limited resources. People visiting the cemetery
> also like the view of the market garden, they find it pleasant, relaxing to
> look at. [The underlying issue is] that councils want the rates that comes
> their way from development. I am very cynical when it comes to councils.

In Sydney, an interview at the Erskineville Community Garden revealed
that development pressures pose a threat to the future of this garden site
(Thornton 2017). What began as a 'guerrilla garden', interest in food
gardening at the Erskineville Road had expanded among the area residents (Thornton 2017: 978):

> [This community garden has] been here for 5 years, just after [the con-
> struction of] one of these buildings. They used this garden site for the
> building projects. Before, this vacant area was used as a car park. We asked
> for a grant for community garden, to enliven[ing] the neighbourhood. But
> Council said no. They really wanted to sell it for development. They said
> our garden did not meet [Council's] criteria for a garden, [as it is sited on]
> a main road, so access is a problem. But [we] could argue there are gates.
> Also, [they say we do not meet the criteria for] security and visual; we need
> two points and they say we have only one. But we could argue that we
> have two-points with all of these windows [from the housing units].

The Erskineville gardener continued to express her disappointment
when working through the Council's criteria for community gardens,
which she found contradictive and, overall, a counterproductive experience. Many city councils in Australia provide information for residents
or groups seeking to create a community garden (through council-de-
termined 'tips or 'checklists'). The process of seeking council information about and developing a community garden plan for council approval
often leads to an exercise in frustration, as follows (Thornton 2017:
978):

> We had a soil test done. There is no contamination, but pollution is pos-
> sible due to continuous traffic. But we grow on top anyways [using raised
> garden beds]. The City [of Sydney] does acknowledge that it is difficult
> to meet all criteria in the inner city. With all [of the] time we spent with

Council to satisfy health and safety requirements we had to make spaces
for access, put tennis balls on garden stakes...they still rejected it.

Council says we love community gardens, but we want to show you a
different site, [suggesting] five alternatives. [There are] lots of little parks
around here, but [they] are not very big. With this site being vacant, it is
not taking anything away from anybody. This site was never given a role, it
was vacant [so we used it for a garden], we enlivened it.

We have had good five years, we would like another five or ten. At least
they [Council] have not done anything with it. It has been zoned 'road',
or there is another word for it. If it were to be sold or built on, it would
have to be re-zoned to residential. [In any case], for a community garden,
it would have to be re-zoned anyway.

Thornton (2017) provides an example of community garden activism bolstered by knowledgeable citizens. In Sydney, Michael Mobbs, is
a local sustainable community activist and author of *Sustainable House*
and *Sustainable Food* (2010, 2012, Choice Books), who rallied community support for growing food on road/street-side verges and setting up
community compost bins. Mobbs has background in environmental law
and experience in working with local government, he was aware that ask-
ing local Council for approval to embark on these activities would be 'a
waste of time' (Thornton 2017: 978). The primary reason being that no
policies had existed, therefore, the Council's default reaction would be
negative, as Mobbs explains (Thornton 2017: 979):

> [For interested community gardeners], it takes over three years and typi-
> cally four to five to get a council approved community garden. There is no
> time limit on council decisions, no statutory goal of supporting or buying
> from local food producers. No councils [in the City of Sydney] buy their
> food from local farmers or community gardens. The Sydney City Council
> is closing one [of the existing community gardens] down against united
> community wishes at Erskineville. [In another community], in Coogee, a
> garden took four years and most [of the] original members had left.

Currently, no city–community participatory frameworks exist for urban
food system policymaking in Australian cities. Although there are some
supportive councils, for example, the City of Melbourne includes community gardening within its wider goals for urban sustainability planning,
as important for public health and well-being. However, as is largely the
case at city-level, approval of urban gardening spaces is subject to how

they might affect neoliberal development plans of a city, as opposed to meeting the needs of the community. For the time being, forms of urban agriculture in Australia, such as community gardens, market gardens, roof gardens and urban farms are only considered in a piecemeal or ad hoc basis (Pires 2011; Thornton 2017). These activities are effectively non-existent (as a holistic strategy) in formal city planning and land use zoning (Pires 2011). In challenging city council for use of urban space—as determined by the community—interview respondents indicated the importance of identifying capacity at the grassroots level (e.g. environmental lawyer) to challenge city officials, and this offers an example of empowered citizens mobilising to claim their 'rights to the city'.

Somerville and Boston, Massachusetts, USA

Somerville and Boston are neighbouring cities where, despite significant urban population densities, zoning for urban agriculture has been completed. As a case in point, the City of Somerville, Massachusetts is ranked 15th nationally in urban population density, which is a measure of the number of residents per square mile (mi^2) (see Table 3.1, with conversions to square kilometre, km^2).

The Boston area is not only characterised as having a mid- to high population density, but as a region, it is experiencing a period of post-industrialism. Using a 'services to goods ratio', Boston is ranked as the 5th most post-industrialised metropolitan area (over one million inhabitants) in the United States as the most densely populated city in New England. Somerville (5 km distance to Boston) was the first city in New England to pass ordinances to adopt urban agriculture in city planning, in 2012,

Table 3.1 Comparison of population density US and Australian cities

City (excl. metro)	Population (2010)	Land area (per mi^2/ km^2)	Population density (per mi^2/km^2)
Boston	645,149	48.43/124.3	13,321/5190
Somerville	75,754	4.11/10.6	18,431/7146
New York	8,175,133	303.31/785.57	26,953/10,406
Sydney	4,627,345	4689.1/12,144.6	986/381
Canberra	381,488 (2013)	314.4/814.2	1213.3/468.5

From compiled sources: City of Sydney; United States Census Bureau; Australia Bureau of Statistics; city-data.com with Boston following soon after (2013). A Somerville city planner stated,

> Over ten years ago, the Mayor's office created the 'Shape-Up Somerville' program, which has become a national model for anti-obesity prevention. Mayor Curtatone was part of a 'Mayors Group' touring urban agriculture sites as part of developing Michelle Obama's 'Let's move program'. In Somerville, a ground swell of young urban farmers, beekeepers and chicken keepers came out to support the ordinance. Also, we a long tradition of Italian, Portuguese and other old-time farmers who have been growing food in Somerville for a very long time. (Somerville city planner 2014, personal communication)

In December 2013, the City of Boston established 'Article 89', which essentially legalised, through a two-year zoning and community consolation process, various forms of urban agriculture in the inner city and wider metropolitan area. In line with the global view of urban agriculture, food production and marketing had long existed in the city, but no frameworks had previously existed to support it. As Popovitch (2014: 1) explains, "with over 40 food truck companies, a pilot residential composting program, 200 community gardens, 100 school gardens and 28 farmer's markets, Boston was in need of a framework for its growing sustainability efforts".

In Boston, the non-profit *Victory Programs* operates an urban farm (*ReVision Urban Farm*), is addressing issues of food and social justice in low-income neighbourhoods in Dorchester (South Boston), where convenient access to fresh food outlets are extremely limited. The lack of availability of fresh produce creates a habituated reliance on foods with low nutritional value, such as highly processed foods and fast foods. In realising that local residents do not "understand vegetables", *Victory* teaches cooking skills to residents ('Rachel' 2013, *ReVision* Urban Farm, personal communication, 17 April). Many of these residents then participate in the urban farming activities, such as developing seedlings and selling produce at local farmer's markets and community supported agriculture. This urban farm is also a pilot project run in partnership with the City of Boston, where the Mayor, Thomas Menino, established a commission to re-zone the inner city and wider metropolitan area, to allow for various forms of urban agriculture. John Read, a senior urban planner, emphasised that urban agriculture is a participatory, proactive response from the city to address issues of social, economic and environ-mental and food justice (John Read 2013, personal communication, 29 May).

These examples are demonstrative of Lefebvre (1991, 1996), Soja (2010) and Harvey (2008), in calling for urban dwellers to assert their 'rights to the city', by carving out just spaces in a post-industrial neolib-eral urban landscape. Urban agriculture is one form of social transforma-tion taking place in cities in the 'global North and South'. Government, or official, support for hitherto grassroots movements for socio-eco-nomic change can be beneficial through removing legislative barriers and providing infrastructure improvements and upgrades. However, there is the hypothetical possibility of government involvement could be viewed as 'interfering' with alternative social spaces that prefer to operate inde-pendently from bureaucracies attempting to regulate 'radical' activi-ties. It could be that official 'take over' of socially re-produced spaces could see them re-emerging elsewhere in the city in reaction to unde-sired bureaucratic influences. This leads to an interesting question of what happens to alternative social movements when they become 'main-stream'? As Crane et al. (2013) suggest, these are questions ideal for crit-ical urban geographers and radical social scientists, to take up the notion of collective urban rights and their potential power.

According to City Farmer, the 1980 US Census revealed that about 30% of America's food, by value, was produced from urban agriculture. The 1990 Census revealed the total to be moving closer to 40%. The United States Department of Agriculture (USDA) does not make a dis-tinction between food produced in rural or urban areas. For the 2017 US Census, urban farmers, who meet the following USDA definition of 'farm', were encouraged to participate in the next survey: "a farm is defined as any place that produced and sold, or normally would have sold, $1000 or more of agricultural products during the Census year".[4]

Urban agriculture appears to have a track record of acceptance at the governmental, non-governmental and local level, as a livelihood and cop-ing strategy across the United States. Where rural farms are declining, urban farms are found to be increasing. A 1993 report from the USDA (Heimlich and Bernard, in Brown et al. 2002: 11), estimated that one third (696,000) of the 2 million farms in the United States are located

---

[4]http://ucanr.edu/sites/UrbanAg/Research/Statistics/.

in metropolitan areas. In this report, three overlapping and broad categories of urban gardeners have been identified: backyard gardeners, community gardeners (local residents growing food for local sale and consumption) and commercial growers, which all contribute to food security and raise the bulk of food involved in urban agriculture (Brown et al. 2002: 12). In many cases, community UA projects have incorporated innovative land use to help solve social problems in urban areas (Thornton 2012, 2017). In Santa Cruz, California, a 2.5-acre lot was turned into a community supported agriculture (CSA, discussed below) project that offers homeless people an opportunity to work and to serve the community.[5] On the east coast, Local Harvest in Massachusetts, maintains a comprehensive directory of over 4000 CSA farms that serves 'tens of thousands of families'.[6]

Often connected with the economic viability of urban agriculture systems are specialised agri-business schemes. For example, in the southern state of Georgia, turf grass (e.g. for athletic competition surfaces) and environmental horticulture are multi-billion dollar agri-businesses. Beyond agri-business, the popularity of community farming, such as CSA, combines the demand for fresh, locally grown produce with idealist small town values and environmental conservation, where farmers are viewed as 'stewards of the land' (Thornton 2012).

London, United Kingdom

An earlier study of urban agriculture in London, UK (Garnett 2001) found that half of Greater London's 2.8 million households have private gardens and long waiting lists for gardening plots or allotments exist, despite the existence of 13,566 ha of farmland on the Greater London fringe (Garnett 2001: 478). However, the availability of land on the peri-urban fringe is diminishing, largely due to development pressures (ibid.). Overall, the outer London plot vacancy figure stands at 81% and 4% for inner London (ibid.). Currently, 30,000 Londoners actively work allotments on 831 ha of land, 111 of which are in inner London. More recent studies have found that privately owned gardens cover 22–27% of the total urban area in the UK (Loram et al. 2007).

---

[5] Homeless Garden Project. http://www.homelessgardenproject.org/programs/.

[6] Local Harvest. https://www.localharvest.org/csa/.

Where allotments have had legal protection since 1908, government and local authorities have been active in promoting and providing for allotments. Local municipalities outside of the city largely own and manage the allotments, though this is less common in inner London, where both the local government and privately owned companies provide for allotments. Within London, the railroad and other companies own 6% of the active allotment sites. Garnett (2001) claims that the number of plot holders under the age of 35 years has been increasing since 1993 and is viewed as a 'youth movement in city gardening'. City gardening was once the preserve of middle-aged residents to supplement the household budget, is popular among London's wealthy young urbanites, which now garden for non-economic reasons. Overall, London's higher income groups (18%) are more likely to grow their own vegetables than lower income groups (11%) (Garnett 2001: 484).

Also, popular in London is the City Farms movement that began in the 1970s.[7] These farms are funded privately and through municipal funds, with livestock being the dominant practice, but they also produce food and ornamental plants. Local communities manage their respective farms, which serve both an educational and community role. Further, the Federation of City Farms and Community Gardens are linked to the majority of London's community gardens.[8] These gardens are located on private or vacant land, as well as in community centres. Established in 1930, the National Society of Allotment and Leisure Gardeners represents and promotes the interests of allotment gardening. Other groups supporting the practice are the Allotments Coalition Trust and Sustain's City Harvest. From a tourism standpoint, these urban and peri-urban farming programmes draw 650,000 visitors per annum, jointly.

Marketing UA: Community Supported Agriculture (CSA)

Community supported agriculture (CSA) promotes and exchanges organic and sub-organic agricultural products through veggie (vegetable) box schemes. CSA requires community membership, where local farms are supported by CSA members placing standing-orders for

---

[7]Throughout the United Kingdom, there are 65 City farms, 8 are located within London and range from 0.25 to 2.5 ha in size (ibid.: 481).

[8]Currently, there are 77 community gardens in London inner-city areas; none of them are commercially viable (ibid.: 487).

produce prior to harvest periods, thereby ensuring the economic survival and viability of local urban, peri-urban and rural farmers. Similar to Von Thünen's land use model,[9] CSA serves as a mechanism in the food supply continuum, which reconnects urban consumers with the value of supporting local small-scale food producers. Similarly, in the UK, Canada and throughout Europe, produce is transferred through local exchange trading schemes—LETS. In Boston, Massachusetts, low-income households can qualify for federal vouchers that are exchangeable for locally grown fruits and vegetables at farmer's markets.

Throughout the developed world, numerous groups, organisations, government agencies and institutions promote, market and transfer organic production and composting household waste. Overall, the holistic allure of community gardens in developed countries is linked to social, health and environmental gains. Emerging and entwined concepts such as new urbanism, smart growth, eco-city (such as Freiburg, Germany in Medearis and Daseking 2012), eco-neighbourhoods and urban greening all emphasise, to various degrees, a growing need to confront the negative spatial impacts of neoliberal urban planning and threats to the global food system ('farm to fork') from climate change. Alternatively, the experiences of the developing world outlined below, indicate urban agriculture's struggle for acceptance, despite the emergence of sustainable urban development in some countries as a key element of social policy for poverty reduction. As described in Opitz et al. (2016: 342), the practice of UA in developing countries is often not a choice, rather, it is a means of survival in providing people not only with food, but also a livelihood.

The existence of a grassroots community activism also serves a type of marketing for a neighbourhood, urban district or for an entire city.

---

[9] In his book Der *isolierte Staat* [The Isolated State] (1875), the German agriculturalist Von Thünen imagined an isolated city set in the middle of a level and uniformly fertile plain without navigable waterways and bounded by a wilderness. He created a model for agricultural urban land use to explain the relationship between the costs of commodity transportation to the location of production. In this model, Von Thünen showed concentric zones of agricultural production. Heavy products, in proportion to value, and perishables would be produced close to the town, while lighter and durable goods were produced on the periphery. As transport costs to the city increased with distance, the returns to the land would diminish until, at a certain distance, land rent would become zero. Moreover, methods of cultivation would vary; with land cultivated more intensively near the city, where it was more valuable.

By 'word of mouth', to social and mainstream media and tourists, cities and the areas in them develop a reputation that attracts the interests of tourists, home-buyers, investors that want to tap into or exploit the draw of urban areas with a 'bohemian' allure in the heart of Berlin. This can result in a backlash by the locals seeking to protect what is unique about their 'place' and the spaces in them. In Berlin-Kreuzberg, the 'bohemian allure' of this district in the former East Berlin had attracted the attention of Google Corp, seeking to build an HQ. This resulted in a swift response from local activists, fearing the homogenising effects of gentrification and significant increases in rental and housing costs. Given the possibility of positive economic returns to the Kreuzberg-Friedrichshain District, investors such as Google Corps will likely win-over local bureaucrats. At some point in the future, locals may turn to new arrivals and visitors in lamenting the past, speaking in nostalgic terms, 'I remember Kreuzberg and its community gardens…before Google came here'.

A CASE STUDY OF URBAN ACTIVISM AND COMMUNITY 'SOCIAL' GARDENS IN BERLIN, GERMANY

This section will address the following questions regarding urban activism and city–community engagement in carving out gardens as 'social spaces' in Berlin—a city experiencing rapid population growth and struggling to maintain and preserve its green open spaces: (1) are there existing frameworks or formal processes for city–community engagement in urban food space issues and (2) to what extent are they replicable to other cities (with due consideration for how geography, culture and history influence places and spaces)? As mentioned in Chapter 1, an objective of this chapter is to discern to what extent UA in cities in developing and developed countries are creating spaces of social and food justice, and related challenges and opportunities for cities to provide and communities to access space. This section will feature recent interviews with urban activists and city officials in Berlin, Germany.

Berlin, Germany

Gardening in the city has a long history in Germany, with the most well-known example being Schrebergärtens (also known as kleingärtens, or 'small gardens'), which are a system of garden allotments, managed at the local level by gardening associations, and have existed for over 150 years. Membership in a kleingärten association are required by law to grow food, loosely understood as 1/3 fruit growing, 1/3 vegetable production, with the remaining 1/3 can be used for non-food gardening. Despite this longevity, even these gardens are 'temporary' in the sense that they do not benefit from any form of permanent land use or zoning status. They have also experienced periods of expansion and decline, due to tensions caused by urban development pressures and the desire of urban dwellers to preserve urban space for food production and well-being. For example, in the immediate years following the end of the Second World War, there were approximately 800,000 allotment gardens in Germany (Gröning 1996), including 200,000 in Berlin (Drescher 2001). Today, there are around 70,000 of these remain in Berlin. If not for the persistence of community groups arguing for the protection of allotment gardening in Berlin, this number could be much lower. In 1993, a few years after the Fall of the Berlin Wall, allotment gardens became a formal part of the urban land use plan for the city (Gröning 1996). At the community level, urban social movements play a critical role in demanding urban space to meet the changing needs and desires of urban dwellers for various social uses, including urban agriculture and its related activities. Discussed below, interview respondents stress that access and availability of urban space for these purposes are hard-won over several years, even decades of fighting off neoliberal development planning to secure or maintain their own 'rights to the city'.

Since the 2000s, community gardens, and other forms of urban agriculture, have been expanding in Berlin and this can be partly explained by the capital city's unique history, culture and important social and economic factors. The Fall of the Berlin Wall in October 1989 not only involved the political and economic reunification of East with West Germany, but it also, on a more social level, reconnected Berliners. The removal of the Berlin Wall revealed vast areas of unused urban space that offered Berliners, especially those on low incomes (ranking among the lowest nationally), opportunities to think and act creatively in re-producing social spaces. According to interviews with Berlin urban activists and city district administrators, prior to 2010, the use of vacant or derelict sites was tolerated by city officials, who viewed the city as stagnant in terms of population growth. Also, vast amounts of urban land were 'inherited' from the former East Berlin, which the city did not have the capacity to develop or maintain. Therefore, the use of this 'new' space was quickly taken up by community groups and 'Guerrilla Gardeners'.

Prinzessinnengärten, Berlin-Kreuzberg

Emerging from the Moritzplatz U-Bahn at the corners of Prinzessinnenstraße, Prinzenstraße and Oranienstraße in Kreuzberg, one is greeted by a wire fence being overtaken by vegetation masking a derelict building site. At first glance, it appears that this section of the former East Berlin has changed little since the Fall of the Berlin Wall in October 1989. A closer look reveals a thriving 6000 m^2 community-led project, the Prinzessinnengärten, which in addition to mobile gardens (developed on crates and rice sacks), it includes a café, restaurant, library, nursery and compost areas. Prior to 2005, the population of Berlin was seen as stagnant. The reunification of the city after the 'fall of the Wall' triggered a wave of outmigration from the city, particularly from Districts in the former East, into the surrounding suburbs and the former West. In 2009, Prinzessinnengärten emerged out of this context, when co-founders Marco Clausen and Robert Shaw, through their Nomadisch Grün not-for-profit (Nomadic Green), began renting an abandoned public lot from the City, which had been identified as a possible motorway junction. Although various gardening activities indeed take place here, Prinzessinnengärten is more than a community garden and, despite its international reputation of locals 'reclaiming the city', the future of this project is very much uncertain (5-year lease expires in December 2018). This uncertainty has little to do with the status of the garden within the community, rather it represents a challenge to the District Administration of Kreuzberg-Friedrichshain (where Prinzessinnengärten is located), in its approach and commitment to community use of vacant spaces and green fields for community-defined purposes.

The following are verbatim excerpts of an interview with Marco Clausen, co-founder of Prinzessinnengärten. As with any piece of qualitative research, there are a number of considerations that factor in deciding on constructing a narrative from interview transcripts. The primary motivation for providing a verbatim account of an interview is to give the respondent a 'voice', while also improving verifiable reliability. It can also illustrate the meanings that people attach to social phenomena (White et al. 2014). In providing an authentic account of the interview, the following contains only minimal edits, where it would improve the 'readability' of the transcript (where German is the first language of the interview respondents) and all clarifications from the transcription process (of audio recordings) are placed in brackets {...}. Of course, there are pitfalls in providing rich, in-depth accounts, in that it can overwhelm a reader with too much detail. Notwithstanding, the complexity of the relationships between city administrators and community activists is best captured in the unfolding 'story' below, in a conversational-style interview with Marco Clausen and a member of the Berlin Senate, Ms. Ursula Renker. In a conversational narrative, the words of the respondent and of the interviewer are viewed as important, as they are engaging in a process of joint production of meaning (Corden and Sainsbury 2006: 11). The narrative will unfold as a 'Q&A' between the interviewer (Alec) and the respondent (Marco). My analysis of the excerpts will feature at the end of each exchange between 'Alec' and 'Marco'. In an effort to check validity through triangulating responses among different interview data sources (Bengtsson 2016), these excerpts will be juxtaposed with responses from an interview with the co-founder of the 'Allmende Kontor' community garden at the disused Berlin Tempelhof Airport, Dr. Elisabeth Meyer-Renschhausen—an urban activist, journalist and academic at Freie Universität Berlin. In doing so, the qualitative research process used here takes a relativist perspective, as it recognises that one point of view is not any more or less valid than another (Barbour 2001). Thusly, this approach acknowledges the existence of multiple views of equal validity. As described by Downe-Wamboldt (1992), the goal is to link the results to their context or to the environment in which they were produced. My analysis will attempt to identify (1) formal processes for city–community engagement in urban food space issues and (2) to what extent are they replicable to other cities.

Interview Results: Urban Governance and City–Community Engagement in Berlin

This section will present interview findings from the first line of inquiry, regarding the existence (or not) of a process for urban governance in Berlin, for city–community engagement in urban food systems and the creation of a 'food policy'. This will be followed by a discussion of the findings related to the second line of enquiry, which explores to what extent is the 'Berlin experience' of urban governance and community engagement replicable to other cities, in countries beyond Germany.

As mentioned above, interview responses to these two main questions will be triangulated throughout the commentary analysis to follow. The substantive interviews were held with Mr. Marco Clausen, co-founder Prinzessinnengärten, and Ms. Ursula Renker, Berlin Senate. I triangulated their responses with Dr. Elisabeth Meyer-Renschhausen, from Freie Universität Berlin (founder Allmende Kontor and AG Kleinlandwirtschaft, or Working Group for Small-Scale Agriculture). The interview respondents were asked to consider the following: 'Is there a process for urban governance in Berlin, for city-community engagement in urban food systems and the creation of a food policy?' In the following excerpts, Marco Clausen follows up his short answer to the question of the existence of a food policy, with a discussion of the broader context of food and agriculture in the EU and in Germany.

Marco Clausen, Co-founder Prinzessinnengärten, Möritzplatz, Berlin-Kreuzberg District

Marco: There is no 'process' {and} there is nothing like a local food policy {in Berlin}. This is something that is just non-existent. 'Food Policy' we only see in the form of {a broader} agricultural policy and is mainly decided on a European or national level {such as the EU Common Agricultural Policy—CAP}. This is important to know that this is the major framework. And the general direction, as seen in the Free Trade Agreement {FTA} discussions, is an open market, globalised food production and consumption and this is fostering a tendency for industrialisation to scaling up agriculture operations. It is worth underlining that {FTA} is not just the United States and Canada, but also EU and West Africa, and other places.

And even though we have around a 10% market share in organic food growing in Germany, but also there you can see heavy industrialisation and huge market power of the huge market chains like ALDI and Lidl {discount supermarket chains}. And you see that small-scale agriculture, or what you might call traditional agriculture, is vanishing, and this has been an on-going process since the 1980s.

With all of this, we have the side effects talked about in different contexts, such as the use of pesticides, air pollution, the use of nitrogen in the soil and water issues, and ethical questions about livestock operations, but it is a disparate field of CSO engagement and activism and policymaking and so on.

But when it comes down to actual food policy making, my understanding of this is, government regulation that refers to the whole process of food production, distributing the field and dealing with the waste {the food supply chain or network}. This is my understanding of food policy, and of course, including health, ecological and economic issues. This I don't see on the national level and this is separated in different fields, like health, economic growth, subsidising farms, so I don't see a coherent policy that {pulls} this all together.

And this is especially true for the ordinary {local} level of municipalities in cities. I cannot talk about all of the German cities, but to my knowledge, this is not a political field that is defined. So, what we see now in the last 2–3 years are tiny initiatives like in Cologne and Berlin are trying to copy this model of Food Councils, basically from North America. For two years now, there is a group called Ernährungsrat. I am not involved with them, but I would say that their main aim is some sort of local lobbying. The political or social movement context of these initiatives is, for example, there is a food movement that is most visible in yearly demonstrations called 'Wir Haben Es Satt' {We Are Fed Up}, and that is a kind of alliance of NGOs, alternative farmers' association that share common ground that they are protesting against industrialisation of agriculture in favour of small-scale farming. It is not just including organic; it's also family-run conventional farms. And even though there are some big environmental protection NGOs, you would see people protesting mass livestock operations, so it is quite a diverse alliance and they bring more than 10,000 people on the streets. And they do this on a symbolic date. In Berlin, we have this 'food fair' or 'farming fair', it's called the Green Week {Grüne Woche}, so it's kind of a protest against it {the industrialisation of agriculture}.

The biggest organisation, when it comes to farming, is Deutsche Bauerverband, a German Farmers association. It's very conventional, conservative and in favour of globalisation and industrialisation.

Alec: Do they look at the organic food movement as a threat?

Marco: Their main role is lobbying for a framework for their core group of farming operations, large, heavily subsidised. They say they represent farmers, but this is a representational issue. They look at small-scale farmers as a marginal thing.

For decades, they have been close to the political system, so when it comes to regulation at the European level, the chemical industry and this type of farming lobbies have huge {influence}. So, the bigger picture is important. There is no local food policy; these are at the national and European levels {on the one hand} and, {on the second hand}, you have NGOs and civil society arguing for a different form of food system.

When it comes to Berlin, when you look at its governance structure, it is very telling about what the food supply for a city is all about. Berlin has no agriculture ministry. As you know, Berlin is a federal county. Other federal counties have ministries that are represented in an assembly of federal states. Berlin decided, in 1990 after the fall of the wall and reunification of the city itself, to give this away to the federal county of Brandenburg. So, actually, even though we are more than 3 million consumers, eating food every day, we are not represented on a national level when it comes to agriculture policy.

When it comes to food policy, we also don't have any senatorial branch. What we have is, only, a department for City Development and Green, it is about parks and waterways, not food. We have a Senate for Justice, and within the Senate, there was a very small department for consumer protection, and the woman that was responsible for this, she was personally interested in food issues, but she was very limited in influence and resources. She was supporting this and she was from the Conservative Party. She is gone now. Let's say for the moment, I have not seen this {food issues} as a main topic for the government. It would be naturally a field for the Green Party, and those who I know in this party are concerned {about this issue}, but I have not read anything about a {food} policy.

Commentary Analysis: Urban Governance and Food Policy in Berlin?

In addressing the first line of enquiry, Marco flatly rejects the notion of a formal process for city–community engagement in urban food systems. An urban food policy for Berlin is also described as 'non-existent'. The purchasing power of Berlin, as a city of 3 million consumers, was raised as an argument for the potential of a regional food policy, but the political will does not exist to make it a key policy concern. Marco made several comments that highlight the primacy of national and EU-level agricultural policy and the influence of conventional or conservative farming associations, such as Deutsche Bauerverband, that are focused on 'the bigger picture' of the globalisation and industrialisation of food supply. This view is consistent with broader, international debates that take a critical view of the globalised food system as contributing to world hunger and poverty (where food distribution is tied to the market—food flows in a cash economy; it is an economically pragmatic system, not philanthropic or humanitarian). It also contributes to social inequality, environmental degradation and is vulnerable to shocks that are climate-related (drought) and market-related (cost of fossil fuels, as it impacts the cost of food production and transportation).

Similar to other cities, globally, these issues are politicising local and regional food issues, including the production of urban food spaces. However, there has yet to emerge a policy framework that pulls the wide ranging—and overlapping—interests of government and civil society on issues including health, nutrition, economic growth, climate change and urban sustainability. In 2015, food policy for Berlin-Brandenburg was discussed and approved by the Senate, under the Administration for Justice and Consumer Protection, but nothing tangible had emerged from this process.[10] Interviews with Ms. Ursula Renker and Dr. Elisabeth Meyer-Renschhausen had indicated part of the reasoning behind this food policy impasse, which was also touched on by Marco. For Ms. Renker, this issue concerns the availability 'cheap food'. It appears that the social and economic costs of lack of engagement in a regional and urban food policy are effectively facilitating, as Marco explained, the

---

[10]This description of the Senate Administration for Justice and Consumer Protection is a Google Translation of the following webpage: https://www.berlin.de/sen/archiv/justv-2011-2016/2015/pressemitteilung.379688.php.

consolidation of organic market power to the discount market chains of ALDI and Lidl. This is having a negative impact on small-scale market farmers in the Brandenburg region, by increasing the buy-in costs for trading in organic produce, which is negatively impacting the rural and peri-urban agri-sector. However, in the following excerpt, Ms. Renker views the market dominance of a few supermarkets in a more positive light, where she rejects the economic value of urban food growing, given the availability of cheap food at the ALDI and Lidl discount chains. She compares her views on social and economic benefits of urban gardening with those of Elisabeth Meyer-Renschhausen (or 'Elisabeth'):

> *Renker:*   For Elisabeth, she thinks that there is an economic reason for people to produce food, if they are low income, they can subsidise it by growing some food. {In my view,} it is more of a social movement because the space is not big enough to really nourish people or families.
>
> *Alec:*   It sounds as though the social aspect would be the 'common ground' shared between the two views of social and economic gardening—growing food brings people closer to nature.
>
> *Renker:*   Yes, we can agree on social aspects of urban gardening, but we disagree when Elisabeth brings in the economic reasons for growing food in the city. People can just buy food cheaply at ALDI or Lidl.

In the above excerpts, Ms. Renker views community gardens as 'social gardens', where they have a social value that can be seen as more important than the food they produce. In particular, as she argues, that there is not enough space in a community garden to really nourish people or families. As a counterpoint, Elisabeth Meyer-Renschhausen argues that urban food growing is economically beneficial, particularly for low-income households, as they can, and do, subsidise the incomes of low-wage earning Berlin households by growing food. Ms. Renker acknowledges that she and Meyer-Renschhausen have a fundamental difference of opinion on this issue,

> *Renker:*   There is a big difference between Elisabeth Meyer-Renschhausen and me, as to why people are doing urban gardening. Elisabeth thinks that it [offers] a big chance to support them economically, because when they can do urban gardening,

they can supplement their income through growing fruit and vegetables, so they do not have to buy so much. Urban gardening is a social aspect of community building, of exchange, of meeting people from all cultures, because gardening itself, seeding, growing and harvesting are things that are in the genes of humanity, and for that reason, people are keen to do this, to keep a kind of village character when producing and gardening. Neighbourhoods are playing an important role because in the city of today, with all of the problems, the gardens are producing a closer neighbourhood and an identity. 'I can meet you there'; it's not in a virtual world, it's very analog and it's very direct. And that's good.

Although there is limited tangible evidence of official city-level support, there is increasing interest at the local level in food policy councils, based on the North American concept that first emerged out of Knoxville-Knox County, Tennessee in 1980. This concept involves a range of stakeholders connected in same way to the food system—at the government, non-government, private business and consumer levels. Although currently small, food policy councils ('Ernährungsrat'[11]) have only recently emerged in German cities, specifically in Cologne and Berlin, and these are driven by non-profit groups. Globally, numerous cities have signed on to the Milan Urban Food Policy pact (2015), including Berlin and Cologne. This has led to urban food movements advocating for 'food councils', also referred to as 'nutrition councils' in Berlin and Cologne. From the Ernährungsrat Berlin website,[12] these councils seek a commitment to the creation of a sustainable food and agriculture policy in the region comprised of a broad alliance of citizens. It aims to actively promote the future viability of the food system in the region. Related to this food policy movement are food alliances, which consist of civil society groups, such as 'Wir Haben Es Satt' (We Are Fed Up) and alternative farmers' associations. These groups share 'common ground' in protesting against industrialisation of agriculture in favour of small-scale farms (organic and conventional, or non-organic). Wir Haben Es Satt is a part

[11]This description of 'Ernährungsrat' is a Google Scholar Translation of the following webpage: http://xn--ernhrungsrat-kln-xnb94a.de/ueber-uns/.

[12]This description of the Ernährungsrat Berlin is a Google Translation of the following webpage: http://ernaehrungsrat-berlin.de/.

of a global movement demonstrating for healthy eating and more rural and ecological farming, animal welfare and fair trade. In January 2017, 18,000 people had gathered in Berlin as part of a demonstration against the industrialised agricultural industry.[13]

In terms of the second line of enquiry, regarding the replicability of the 'Berlin experience' of urban governance and community engagement, Marco emphasised that Berlin has a long history of social movements that may give the appearance of strength and success in producing urban social space. The reality behind any 'success', in carving out and holding onto urban social space, was preceded by years or decades of communities and urban dwellers 'fighting' for their rights to the city. The strength of social movements is dependent on the depths of communities politicising and becoming directly engaged in local issues—urban food, affordable housing, jobs or environmental issues. As explained by each interview respondent, the use of Tempelhof Airport, the disused airport, as a public open space and community garden (Allmende Kontor) was the product of local people becoming politicised, informed and organised to a level where they could challenge the City of Berlin, and hold it accountable to the principle of community/ citizen engagement in planning decisions regarding the use of vacant, or disused, public space (the 'Tempelhof Field Law' is discussed, below). Although Marco is sceptical about the future of the Tempelhof Airport remaining a public open space, Ms. Renker and Dr. Meyer-Renchhausen were confident in stating that Tempelhof Field will not be developed.

In establishing the 'bigger picture', Marco's responses demonstrate the depth of understanding and concern that local urbanites have of global issues and its local level effects. His comments also reveal a level of frustration with city engagement and uncertainty in the future of community-led 'start-ups' such as the Prinzessinnengärten. In the following excerpts, Marco continues the discussion of food policy and the extent of community struggle in claiming their rights to the city. This discussion also directly engages with the second line of inquiry, to what extent is the Berlin experience in urban governance and community engagement replicable. For this study, the potential for 'replicability' relies on the emergence of politicised groups or communities that hold their governments accountable to notions of transparency in decision-making on the

---

[13]Agriculture and Rural Convention 2020 (ARC2020): http://www.arc2020.eu/ soup-talk-2017-sharing-ideas-inspiring-action/.

use of public, disused and vacant space. It is the politicisation of urban food space and, more broadly, of people exercising their 'rights to the city' that Marco continues to emphasise, in the following excerpt, as a strength of Berlin social movements:

Marco: Even though the City of Berlin itself is praising {urban gardening}, they do nothing to make this {urban food policy} happen. All of the community gardens are done by citizens' initiatives, with their own energy and resources, all through heavy conflicts about land access and so on. So, every garden that you see has gone through some heavy fighting to exist.

Slowly you see some places, where you have the feeling that the policy might change, but these are exceptions. So, for me, one of the most interesting {exceptions}, and not advertised so broadly, is in the new park at Gleisdreieck, there is a community garden within that park. The park itself has a 30-year history of fights—it's Berlin, like the airport, things take time, but it can be good. In this case, they allowed, as a total exception, a community garden within a public park. Because the standard answer of the administration is 'a community garden is the privatisation of public lands so we cannot allow it. We specifically cannot allow fences and privately managed plots.' It is an interesting example, but I also think they are afraid to create precedent, so that is why you have not seen anything that comes close to a plan for urban or community gardens or UA—there is nothing like this.

There was a study being done by the Senate, about new forms of green spaces in 2001 or 2011, especially stresses places like this of a space that is green, that is social and managed by citizens, but this is sitting on the shelf. There would be lots of things {in it} that I could agree with. There was a master plan for the city, for the Green and the Blue {development}, to find models for contemporary forms of management. But from what I can see, they do not make any use of it {the study}. It went to the parliament, the senate they put their signatures on it {then nothing becomes of it}. There is no idea for UA planning into green and social infrastructure. We don't have a study about how this can fit into the German law or administrative system.

Alec: Outside of Germany, Berlin has a reputation as being particularly progressive in community gardens. Why is that?

Marco: But it is very true that, in Berlin, we are very rich in people coming together to do projects and initiatives, but very poor when it comes to taking these ideas (of UA planning) into the political decision-making process, and find regulations and budgets.

You can find people within the administration, like Frau Renker {Senate}, who are personally invested, and they are nice people, and they also, I have to stress this, they also know their stuff. But they are also exceptions within the administration. And they often do not have the back-up of the rest of the department; it's not considered extremely important, no budget and so on.

And then, it is interesting that in last year's elections, there were posters from different Parties that all showed community gardens. Like the Green Party and Die Linke. And they announced in their coalition contract that they want to support urban gardens, but there is no serious attempt; on the contrary, most of these gardens like this are based on temporary uses, we can talk more on this later, but they are very fragile and based on voluntary work. So, in a way, we do a public service, as these gardens provide ecological and social value and are educational. But it's all based on voluntary work; there is no funding whatsoever.

At Tempelhof Airport, they have a system where people have to pay for using the public space, $5000 Euro. And when you talk about governance, you have hidden privatisation. We see this in Berlin, for the last 10 years, a lot of public infrastructure and publically owned housing was privatised to the highest bidder with all of the consequences. A public housing group was sold to Goldman Sachs, [...], and we are seeing the consequences of that {rising costs}. But what is also happening in Germany, since the 1990s, we are shifting legal entities to private governance models. So, services like the railways that were formerly public were changed to a stock market company that is owned by the state, but managed like a AG, but the same is true for green areas in Berlin and Cologne, we created entities that are limited to do these public services. Grün Berlin is a company that is managing all of the prestigious parks, such as Tempelhof. They are a Limited, with a CEO, so they are not answering 'us', as constituents. They often put fences around green areas, such as Tempelhof, or take a fee to enter them and they get the most interesting parts of the city.

Alec: There are a lot of groups listed under Grün Berlin.

Marco: Grün Berlin are managing the parks; they are in charge. And they have to abide by the law that was made, the referendum. The City of Berlin made a plan to develop the city, without any public consultation. So, the people made a referendum and they won it, for the whole city. And made it clear that this type of top-down planning is not seen as democratic.

Within the referendum you have to formulate a law that people vote on. During the campaign to keep Tempelhof 100% free of development, the initiative concentrated on a rule that does not allow to change anything on the field. What they did not include, as it was only 12 people, and they were overwhelmed, is that they did not include new rules for governance of that space. The 'law' just said that you 'cannot do anything with it' {with Tempelhof}, but what does that mean exactly, to keep it as this kind of park?

–23.00–

Alec: I did ask Frau Renker if the city will leave Tempelhof as an open public space, to not develop it, despite the housing crisis and the city being bankrupt (her response was that it would remain an open space).

Marco: She has to say this; she is part of the administration; she is not a politician. She has a duty, and is very limited in what she can say. They had used the refugee crisis to try to 'test' the law that was not so professionally done, so they failed. The plan they had to put refugees there was absurd. But they will attack it {again}, sooner or later.

COMMENTARY ANALYSIS: REFERENDUM ON THE OPEN USE OF TEMPELHOF AIRPORT

The referendum that Marco is referring to is the Tempelhof Field Law (ThFG). From the City of Berlin's website for the Senate Administration for Environment, Transport and Climate Protection, the ThFG is explained, as follows:

Following the successful decision in May 2014, the Tempelhof Field Law (ThFG) came into force in June 2014. [...] the ThFG provides that a development and care plan (EPP) with the participation of the population must be drawn up for the future use and further development of the open space of the Tempelhofer field. This EPP will include the needs of the protection, conservation, preservation and nursing purposes defined in the ThFG as well as development goals, regulate and organize the nature-relevant, urban climate and landscape design and will serve as the basis for the implementation of the leisure activities and the utilization demands of the population.

The Senate administration for the environment, transport and climate protection wants to implement this statutory task of the people's movement quickly. In order to enable an open, transparent and efficient conception, a broad involvement of the actors of the civil society, the members of the current user advisory council, sports and nature protection associations, small garden colonies, politics, administration and all interested Berliners.[14]

In addition to the unique decision to re-purpose Tempelhof Airport as a public open space, the following excerpts relate to what extent the 'Berlin experience' of urban food governance and community engagement are replicable to other cities, in Germany and globally. Ultimately, the 'experience' of cities with urban agriculture—as a holistic system encompassing the entire food supply chain, from production to waste—is very much place-based. This is not to say that important lessons cannot be shared; to the contrary, the experiences of others—and of other places in the re-use of urban space—can provide inspiration and act as a catalyst for local action. After all, the founders of the Prinzessinnengärten were initially inspired by the urban agriculture and urban food governance experiences of a country in the global South—Cuba. The following excerpt explores the grassroots perception that the era of social production of urban open space for community gardens, and other community-defined uses, is likely over:

Alec: At the district level, is there any political will to see community gardens, such as Prinzessinnengärten, continue, or

---

[14]This description of the ThFG is a Google Translation of the following webpage: http://www.berlin.de/senuvk/umwelt/stadtgruen/tempelhofer_feld/index.shtml.

would they prefer to 'wish them away' as a form of public use of green space?

Marco: No, they tolerate them, this is a long story in Berlin, it has a long tradition from 20 to 30 years. The way things worked, probably to the benefit of both sides. The administration did not have to deal with it {regulating green/public spaces} and people could just do what they want. And it was a strength, in a way, of Berlin of, let's say 'sloppiness' to allow people to stat gather a start a club, a gallery, a bar, or whatever and circumventing heavy regulation. And in some boroughs or neighbourhoods, what is important is not the {official} support for what you want to do, but what is important is that people {locals} are not against what you want to do. In Kreuzberg, there is a high degree of tolerance where people say 'I do not care what you are doing over there'. You would not describe this as the dominant attitude of Germany, in general.

The following excerpt effectively refutes the notion of replicability of the Berlin experience in community gardens, and enhances the differential nature of urban agriculture and the role that local culture plays in the emergence of urban food space activism, and forms of UA.

Marco: Today, the economy and social texture is changing, so this is not possible anymore ('do what you want with open space') and we do not have the free spaces as everything is 'financialised' and marketed and without any political support. Also, the private lives of people are more and more under economic stress, with less free time than before, and must take on more 'social service' type roles, such as child day care. We still have very low income in Berlin, along with constantly rising costs for not just flats, but public services.

So, overtime, I think we 'drained out' this reservoir of this kind of {loose terms} engagement, so we will see less spontaneous popping up of projects, at least not when they are related to {public use of} space. There is still a lot of things going, lots of young people trying things out, but this {Prinzessinnengärten} type of physical, visible 'taking over' of concrete spaces and doing things on your own terms, which defined a lot of this alternative use of space and urban landscape in Berlin, and I think that this is history. So now, we either step into building reliable economic political frameworks to allow this, or it will disappear over time.

And when it comes to food, I really need to stress this, for me, this urban community garden that we see here has absolutely nothing to do with urban food production. It is dangerous that they market this or publish this as a 'model for future food', or whatever. It is only related to food in terms of creating an awareness, developing practical skills and changing peoples' attitudes and broadened their understanding of what is going on in the world. And that is something that they actually, people become politicised by doing this. When it comes to biodiversity, industrialisation and so on, they have a much better understanding of these because they start dealing with it. So, they do become politicised {once they understand the global and local issues}, but what they don't do is produce a lot of food, and I would stress, at all. It may be a by-product.

You only see some self-sufficiency in some migrant communities. Everything else related to self-sufficiency has disappeared, even in the biggest branch of these urban gardens, the allotment Schrebergärtens. There, you have to grow food, but they are not about this anymore.

Alec: Another interviewee [at the Ton Steine Gärten am Mariannenplatz, Kruezberg] had said that Schrebergärtens are 'holy' in Germany.

Marco: Now, they are fighting also. There is a Federal law to protect them, but they are not part of the planning. They are in a different form of governance. From the very beginning, they were set on places that were designated for future development. Almost all of them. The regulations said that they must grow food, form an association, but it also defines rules for what happens if the city wants to use the land. So, there is a procedure for getting rid of them. The ones in the inner city are under pressure, some are already gone. They have the problem that, since the 1950s, self-sufficiency is not an issue any more, in West Germany, though it was for a longer time in East Germany, but since the 1990s, food became so cheap for people to bother to grow it themselves. There are also the roles of reproductive work 'women's work' and so on; since then, these gardens had transformed into very private, recreational things often combined, with houses, a typical German thing.

Some of them people are living in the houses, parking their cares, trimmed lawns, using Round-up. They have German white males dominating the associations and their ideas of aesthetics and rules. This is a bit of a stereotype, but there is an issue that the never renewed the idea {behind the allotment garden}, which can become a problem in 10–20 years, where they can become a target for urban development, because their protection is not really strong in Berlin, with some only protected until 2020, or so. They can become a target, if they do not reconnect to the earlier ideas of ecological and social issues. But then they'll have to open the whole thing to new forms of management and new functions and so on.

In the following excerpt, Marco discusses the important role that the city could play in supporting a local or regional food system.

Marco: So, we see a developing food movement in the center of Berlin, combined with more wealthy people, tourists, with hipsterisation. And in general, more awareness of organic products, crafted (beer, bread) products. There are resources, when it comes to spaces, knowledge and money to change our relation to the countryside, which for me is the key question. So, we could use this type of {global} interest and buying power to foster a different agriculture in the region. And people in the city love to buy regionally, locally and even the big supermarket chains realise this. What rules our thinking is a blind spot, if you go to {the farms in} Brandenburg itself, there is no small-scale ecological farming, at all. So, we will never be able to {supply it}, if there is a demand in central Berlin for this type of food production, it does not exist. There is an ecological desert out there, where they grow subsidised corn for so-called eco-fuels, where we have mass livestock operations, extremely low wages, privatisation of formerly public farming land. Even if there is a new generation of young people wanting to do agriculture, they don't have access to credit, land and no distribution system whatsoever.

In our kitchen, we try to support a few small family farms working under difficult circumstances, we try to support them. There is no distribution system, so the farmer with his wife, has to spend three days per week to compensate for this {transport food}. If he would enter the existing {mainstream} distribution system for organic food, he would enter this whole world of monopolisation with a lot of pressure on the producers when it comes to price, and what the food should look like, what to grow and so on. They do not earn much and of you have only 10 hectares, you cannot survive. So, what you see are some niches, some CSAs {community supported agriculture}, but they are an exception. You see some mid- to big sized ecological operations some of them do quite good on the market, as we have 60 or so bio-shops. So, if you have this type of size as organic and then you relate it to the supermarkets then you might have a chance, but the overall picture {of the countryside} is still plantations of pine trees that are ecological deserts, or for the corn, on degraded soil. Degradation is a huge problem, and in this part of Europe is probably the biggest impact of climate change. We don't have any policy connecting the market to the producers. Even though now would be a good time to channel resources into this.

For a political decision-making, they should create clusters in Brandenburg, of groups of alternative farms, with a big enough scale so they can set up their own distribution system.

Alec: It sounds like CSA.

Marco: Yes, but bigger. The problem is the CSAs are islands and they are often quite dispersed. For clusters, they would have to be close enough to share machinery and distribution system. This would have to be supported by the City of Berlin using its buying power, its institutions like schools, hospitals and so on, to start this. Because the main problem next to growing on the countryside, is we are losing the culture, the social texture, with the political outcome that everyone is surprised now about, in the West 'why is the countryside betraying us?': It is because we left them alone for the last 40 years. So this is key, that this is not just a food issue, that this is an overall issue of culture and social relations between urban and the rural.

INTERVIEW WITH URSULA RENKER

Senatsverwaltung für Umwelt, Verkehr und Klimaschutz, Berlin
(Senate Department for the Environment, Transport and Climate Protection Open Space and Park Landscape)

In the following excerpts, Ms. Renker establishes her views of urban gardening as 'social gardening' and that these activities are grassroots attempts to 'reclaiming the city'. Ms. Renker's comments on the social importance of urban or community gardens to society, reflect Marco Clausen's view of social impacts/community social value of these activities as a more strategically important position for community garden groups to make, in arguing for urban food space.

Alec: Can you talk about changes you have seen in urban food growing in Berlin over the years?

Renker: When you talk about urban farming, or urban food growing, you have to talk about the allotment culture—the kleingärtens (allotment or Schrebergärtens). We have 70,000 allotments in Berlin. They have a special kind of regulation, where they can rent out space cheaply when they are producing food. Without food production, there are no cheap rates.

The development of Berlin and the culture of urban agriculture, is dependent on the fact that there are official allotments, clearly defined on the planning maps. Allotments are a specific space, especially for that use (for food growing). And then you also have unofficial allotments, called Graberland, where you have informal gardening that is not included in the maps.

There is a third difference; you have the official gardens, which are without any kind of timeframe, and today, in the City's construction and development plan Flächenentwicklungsplan (Area Development Plan) where you have the use of space and the use of temporary spaces, as long as they are not needed they can become allotments, but they are subject to a timeframe.

Alec: If they are not needed for…economic development?

Renker: Yes, economic development, but mostly housing. If you look on the greenfield map of Berlin, you can see a lot of the allotments and, at the moment, they are green, but moving into a strong conservatism.

Alec: What does that mean?

Renker: First, you have the allotment people who are conservative, traditional older people.

Second, you have the trendy, hip people who want 'social gardening' wherever it is possible. Wherever there is a free space, they start a social garden. The urban garden movement is an attempt to 'reclaim the city', so it is becoming partially politicised, but not all of them. The allotment gardens are the traditionalists, but the social urban gardens, such as the Allmende Kontor, they are motivated partially politically, by Die Linke and the Green Party.

Alec: I've heard about the 'reclaim the commons movement.'

Renker: Yes, the Allmende Kontor garden, Allmende is an old German word for 'the commons'. A typical signature for urban gardening in Berlin is that it is starting where there is very dense housing. Where there is intensive housing and very little public space, people get involved in greening any existing space and reclaiming it. And this is the opposite to the outside of Berlin, like Spandau and Falkensee, where you have so much free, public space to do that (garden), where you can receive the space freely (the space is readily available) and quite cheaply. In the inner city, such as Kreuzberg, where you have the old workers' areas, small flats, tiny streets, so it is a counter movement in a way.

PIECE OF GROUND, FOR 'SOCIAL PEACE'

In the following excerpt from Ms. Renker, she explains the 'mood' of city planners in their approach to the vast amounts of vacant and disused space that the City of Berlin had inherited with reunification, after the fall of the Berlin Wall. Entering the mid-2000s, the attitude of the City was that the population was stagnant and there were no immediate development plans for these 'new' spaces, nor did it have the financial resources to manage them. Therefore, any community appropriation of these urban spaces was often viewed in a positive light by city planners and councils through conceptualising the use of these spaces as 'temporary uses', from a planning perspective. Prior to 2005, The Berlin Senate produced a book 'Urban Pioneers' (published in 2007)[15] to both

[15] Senatsverwaltung für Stadtentwicklung, and Denton, J. 2007. *Urban pioneers: Temporary use and urban development in Berlin*. Jovis: Berlin.

promote the notion of 'Berlin as a laboratory' for temporary use and the creativity of Berliners in 'pioneering' diverse forms of 'unplanned' land use in often disparate urban spaces. By 2010, the reality of 'more people moving in, than moving out' of the City had settled in and the need for urban space for housing development had tempered any city-level enthusiasm for social use of Berlin's riches of disused space.

> *Alec:*   As cities become denser, how do you balance the need for housing/economic development (economic use value) with social needs for space (social use value)?
>
> *Renker:* Social peace in the city is depending on this balance, as new studies are showing on urban development, are dependent on the possibility of human beings interacting and meeting.
>
> The development of Berlin up to 2005, was not facing any intensive development, it was more of a decline or stagnation of inhabitants. Since 2005, it has changed quite fast and the need for housing grew over proportionately.
>
> This book Urban Pioneers, was produced before 2005, suggesting {open} field spaces for temporary use, including for agricultural use. In 2005, there was some indication that more people are moving in than out.
>
> In 2010, the governmental structures only officially accepted that change was happening and started to react. People were moving into the older quarters, in the old flats or housing of the last century. People moving in from all over the world became a big topic. They are coming in as part of a 'start-up' scene, American music scene. Turkish communities, with existing families have extending families coming, Arabic communities from war-torn areas.
>
> *Renker:* So, what has changed in food production in Berlin? Nothing.
>
> Berlin has the official allotments, which had already been in the city and produce food. In the City, there are around 90–100 urban gardening initiatives that are doing gardening projects, but compared to the 70,000 allotments, they produce just a tiny bit. But the idea today, is that people want their garden directly outside, where they live, or on their balcony, or at the workplace. There is a change in this mentality, they want to have it 'reachable'.

On the other hand, you have Brandenburg around Berlin, where there is so much space for agriculture but there is the problem that no one wants to go there, to live there, to do it {farm}.

Renker: Urban gardening is a social aspect of community building, of exchange, of meeting people from all cultures, because gardening itself, seeding, growing and harvesting are things that are in the genes of humanity, and for that reason, people are keen to do this, to keep a kind of village character when producing and gardening. Neighbourhoods are playing an important role, because in the city of today, with all of the problems, the gardens are producing a closer neighbourhood and an identity {this can be seen as more important than the food they produce}. "I can meet you there" It's not in a virtual world that's very analogue and it's very direct. And that's good.

It is also something that people can see things through, from the beginning to the end. You seed it, you take care of it and then you can harvest it. And this is something that is close to the 'do it yourself' culture—you can do books together, knit or other things. This is all creating the kind of feeling that 'hey, in a city, you have separated tasks, you only see the part that you are doing, so you never have this feeling of self-fulfilment'.

(When I) was born in the 1955, schools had all kinds of 'do it yourself' projects. This does not exist in schools today, but in its place, you have this 'do it yourself' culture, and it is coming from America, very trendy. But 'Selbstgemacht'[self-made] is German, it is very old.

Alec: Do you directly engage with urban/community garden groups, such as Allmende Kontor?

Renker: Here in the Senate, formally the Ministry for Green culture, climate and traffic, we have 'round table' meeting 4 × per year for the urban gardeners, and all stakeholders are invited. They are normally speaking about problems and conflicts, but recently they are creating a map, which would be interactive for all urban gardening projects in Berlin. The city of Essen just produced such a new map.

Alec: I am very much interested in this last point, where the local or district government meets with urban gardeners 4 × p/a. This sort of urban governance does not happen in

Australia, where UA and the people engaged in it are not taken seriously.

Renker: Berlin has 12 districts, and green fields are the responsibility of these districts. All of them are open, green space. And all of them are handling urban gardening differently.

Most of the urban gardening in the districts are not taking place in the open green fields, but in the private spaces of companies, like the energy, water and transport providers—such as the Deutsche Bahn (e.g. Gleisdreieck Park). They have space that is unused, or no longer in use, and there, people are starting gardening projects. So, this is a different kind of responsibility, as they are not part of districts' green fields.

Alec: In the case of people doing gardening projects on the grounds of private companies, is this a problem for the city?

Renker: No. Not at all. Although a lot of these projects are private {in the sense that they are not taking place on District public green fields}, it was still seen as necessary to support them. The round table, for example, was working on a contract outlining gardening plans or frameworks to take to either a District municipality or a private company, to ask for space to start a garden.

Alec: In using private company lands instead of public green space for gardening projects, does this form of privatisation mean that city districts are divesting their interest or responsibility to provide public space for people to use for various projects?

Renker: The possibility of a private initiative starting green projects in public space is still possible, for example, of they want 5000 m² for a garden project. But the issue would then be that the park is a public space which is paid by taxes. So, if you take 5000 m² away from public use for these groups, then it is a kind of privatisation of public land. So, this something that Districts do not want to support. Increasingly, people are going to private companies to use their unused space. This is not seen as any kind of competition with the District green fields and it does not mean that Districts are lowering their responsibility in providing or maintaining green field activities and support for them.

Urban gardening projects in green parks must be as open and accessible to the public as possible {no fences}. Gurlitzer, Mauer or Gleisdreieck Park, they are just testing that out. These parks also have a big children's component.

Alec: For private companies supporting UA projects at these parks, who are they doing it for? Are they doing it for a public group thus making them defacto public, not private?

Renker: Both.

Alec: Is this an example of privatisation of public land, where the city is starting to remove itself from its responsibilities in managing its green fields?

Renker: They all start with the community idea. In the process, people are identifying themselves with the plant; whatever they are growing {and where they grow it}, it becomes theirs {thus an issue of private ownership}. The feelings and emotions in that process are very archaic, a kind of settlement of humankind. You start to settle, you seed and you grow things. And this is the kind of transformation that people who start gardening are going through. {It starts} from the community idea to 'Hey, this is mine'.

DIMENSIONS OF FOOD SECURITY IN BERLIN

The following line of questioning with Frau Renker gives some indication of how the Berlin Senate views 'food security' as a concept in Berlin. Clearly, the Senate sees urban food production as playing an important educational role. This reflects the fourth dimension of food security, as defined by the FAO (2015).

Alec: Is locally produced food under consideration by the Senate, as a way to enhance food security in Berlin?

Renker: This is still not a topic in the Administration, because we are surrounded by Brandenburg {Berlin is a city-state located in the middle of Brandenburg, which has traditionally been an agricultural state}. {So} producing locally or directly where it is needed is not something that the government is thinking about.

Alec: Food security, is it important, on the social aspect, learning how to grow food? Is this knowledge worth preserving in the city?

Renker: The food knowledge side {of food security}, this is something that would be important to the Administration. The environmental, ecological, nutritional side, but less on the {inner city} food distribution side.

Commentary Analysis: Lesson Learned from Berlin

This section explored Berlin case studies of community gardens and public open space. It addressed the following lines of inquiry: (1) are there existing frameworks or formal processes for city–community engagement in urban food space issues and (2) to what extent are they replicable to other cities? The possibility of open lines of communication between city–community actors is critical for engagement on issues concerning socio-spatial justice. Though, this possibility exists only through the persistence of urban activists 'on the ground'. The potential for 'replicability' relies on the emergence of a politically mobilised citizenry and, as seen in the Berlin example, for demanding transparency in government decision-making on the use of public, disused and vacant space. Interview responses made clear statements regarding past and current examples of the politicisation of urban food space and of Berliners claiming their 'rights to the city'. It was also explained that the claiming urban social space, such as Allmende Kontor and Prinzessinnengärten, were the result of years, even decades, of hard-fought gains by community groups 'fighting' through tensions with city government.

The interviews highlighted the importance of a culture of urban activism, where the initial interest of individuals in, for example, a community garden plot, which can lead to their engagement, at a local level, in global issues. This cross-fertilisation of knowledge can be an effective form of community education and awareness, and has a politicising effect. To the extent that this politicising effect develops into a culture of urban activism is critical, as it anchors the 'staying power' of community or social groups to endure years or decades of negotiations and 'fighting', with city government for their collective and locally determined rights to the city. This development or evolution of an urban activist culture is not easily replicable in urban societies in North or South cities. For the latter, the emergence of an organised and educated urban activism is further challenged by the experience of developing countries in social and political fragmentation, civil unrest and conflict. For these reasons (though not exclusively), community groups can struggle to emerge as a politicised social movement, to seize an opportunity for political engagement, and to shape their relationships with city government and claim their rights to the city (Mitlin and Mogaladi 2013).

REFERENCES

Armar-Klemesu, M., and D. Maxwell. 2000. Accra: Urban agriculture as an asset strategy, supplementing income and diets. *Growing cities, growing food: Urban agriculture on the policy agenda*, 203–208.

Australian City Farms and Community Gardens Network. 2011. https://communitygarden.org.au/2011/12/21/sydney-city-community-gardens-attract-visitors/.

Australian Government. 2013. The state of Australian cities. https://infrastructure.gov.au/infrastructure/pab/soac/index.aspx. Accessed 22 Jan 2016.

Badami, M.G., and N. Ramankutty. 2015. Urban agriculture and food security: A critique based on an assessment of urban land constraints. *Global Food Security* 4: 8–15.

Barbour, R.S. 2001. Checklists for improving rigour in qualitative research: A case of the tail wagging the dog? *BMJ: British Medical Journal* 322 (7294): 1115–1117.

Bengtsson, M. 2016. How to plan and perform a qualitative study using content analysis. *NursingPlus Open* 2: 8–14.

Brown, K., M. Bailkey, A. Meares-Cohen, J. Nasr, J. Smit, T. Buchanan, and P. Mann. 2002. Urban agriculture and community food security in the United States: Farming from the city center to the urban fringe. Available at: http://www.foodsecurity.org/PrimerCFSCUAC.pdf. Accessed 20 Dec 2017.

Bundy, C. 1979. *The rise and fall of the South African peasantry*. London: Heinemann.

Companioni, N., Y. Hernandez, E. Paez, and C. Murphy. 2002. The growth of urban Agriculture. In *Sustainable agriculture and resistance: Transforming food production in Cuba*, ed. F. Funes, L. Garcia, M. Bourque, and P. Rosset. Oakland, CA: Food First.

Corden, A., and R. Sainsbury. 2006. *Using verbatim quotations in reporting qualitative social research: Researchers' views*, 11–14. York: University of York.

Crane, A., L. Viswanathan, and G. Whitelaw. 2013. Sustainability through intervention: a case study of guerrilla gardening in Kingston, Ontario. *Local Environment: The International Journal of Justice and Sustainability* 18 (1): 71–90.

Dasso, A., and T. Pinzas. 2000. NGO experiences in Lima targeting the urban poor through urban agriculture. In *Growing cities, growing food-urban agriculture on the policy agenda*, ed. N. Bakker, M. Dubbeling, S. Guendel, U. Sabel-Koschella, and H. de Zeeuw, 349–362. Eurasburg, Germany: DSE.

De Haen, H. 2002. Enhancing the contribution of urban agriculture to food security. Urban Agriculture Magazine. Special issue, *For the world food summit: Five years later.* The Netherlands, Resource Centre for Urban Agriculture (RUAF).

Dieleman, H. 2017. Urban agriculture in Mexico City; balancing between ecological, economic, social and symbolic value. *Journal of Cleaner Production* 163: S156–S163.

Downe-Wamboldt, D. 1992. Content analysis: Method, applications and issues. *Health Care for Women International* 13: 313–321.

Drescher, A.W. 2001. The German allotment gardens-a model for poverty alleviation and food security in Southern African Cities. In *Proceedings of the sub-regional expert meeting on urban horticulture, Stellenbosch, South Africa,* 159–167.

Egziabher, A. 1994. Urban farming, co-operatives and the urban poor in Addis Ababa. In *Cities feeding people: An examination of urban agriculture in East Africa,* ed. A. Egziabher, 67–84. Ottawa: International Development Research Centre.

Fan, S., L. Zhang, and X. Zhang. 2004. Reforms, investment, and poverty in rural China. *Economic Development and Cultural Change* 52 (2): 395–421.

Foeken, D., M. Sofer, and M. Mlozi. 2004. *Urban agriculture in Tanzania: Issues of sustainability.* African Studies Centre Research Report 75. Leiden.

Food and Agricultural Organisation (FAO) of the United Nations. 2015. *The state of food insecurity in the world (SOFI).* Rome: Food and Agriculture Organization of the United Nations (FAO).

Freeman, D.B. 1991. *City of farmers: Informal urban agriculture in the open spaces of Nairobi, Kenya.* McGill-Queen's Press-MQUP.

Freestone, R. 2014. Australian environmental planning, origin and theories. In *Australian environmental planning: Challenges and future prospects,* ed. J. Byrne, N. Sipe, and J. Dodson, 21–35. New York: Routledge.

Garnett, T. 2001. Urban agriculture in London: Rethinking our food economy. In *Growing cities, growing food: Urban agriculture on the policy agenda,* ed. N. Bakker, M. Dubbeling, S. Gündel, U. Sabel-Koschella, and H. De Zeeuw, 477–500. Feldafing: DSE/Zentralstelle für Ernährung und Landwirtschaft.

Gaynor, A. 2006. *Harvest of the suburbs: An environmental history of growing food in Australian cities.* Crawley: University of Western Australia Press.

Gertel, J., and S. Samir. 2000. Cairo: Urban agriculture and visions for a 'modern'city. In *Growing cities, growing food: Urban agriculture on the policy agenda,* 209–234. Feldafing: Deutsche Stiftung für Internationale Entwicklung (DSE), Zentralstelle für Ernährung und Landwirtschaft.

Gibson-Graham, J.K. 2008. Diverse economies: Performative practices for 'other worlds'. *Progress in Human Geography* 32 (5): 613–632.

Gogwana, M. 2001. The role of urban agriculture in food security: A case of low-income dwellers in Dangamvura. In 2nd WARFSA/WaterNet

Symposium: Integrated Water Resource Management: Theories, Practices, Cases, 30–31 October, Cape Town.

Gröning, G. 1996. Politics of community gardening in Germany. Available at: http://www.cityfarmer.org/german99.html. Accessed 27 June 2017.

Gurran, N. 2011. *Australian urban land use planning: Principles, systems and practice.* Sydney: Sydney University Press.

Hallett, S., L. Hoagland, E. Toner, T.M. Gradziel, C.A. Mitchell, and A.L. Whipkey. 2016. Urban agriculture: Environmental, economic, and social perspectives. *Horticultural Reviews* 65–120.

Hamilton, A.J., K. Burry, H.F. Mok, S.F. Barker, J.R. Grove, and V.G. Williamson. 2014. Give peas a chance? Urban agriculture in developing countries. A review. *Agronomy for Sustainable Development* 34 (1): 45–73.

Harvey, D. 2008. The right to the city. *The New Left Review* 53: 23–40.

Hungwe, C. 2007. The Effectiveness of Urban Agriculture as a Survival Strategy among Gweru Urban Farmers in Zimbabwe. *Urban Agriculture Magazine No. 18-Building Communities through Urban Agriculture*, The Netherlands, RUAF.

Ibrahim, A.A., and S.A. Elariane. 2018. Feasibility tools for urban animal husbandry in cities: Case of greater Cairo. *Urban Research & Practice* 11 (2): 111–138.

Jacobi, P., J. Amend, and S. Kiango. 2000. Urban agriculture in Dar es Salaam: Providing an indispensable part of the diet. *Growing cities, growing food: Urban agriculture on the policy agenda*, 257–283.

Kreinecker, P. 2000. La Paz: Urban agriculture in harsh ecological conditions. In *growing cities, growing Food*, ed. N. Bakker, M. Dubbeling, S. Güendel, U. Sabel-Koschella, and H. de Zeeuw, 391–412. Feldafing: DSE.

Lee-Smith, D., and P. Ali Memon. 1994. Urban agriculture in Kenya. In *Cities feeding people: An examination of urban agriculture in East Africa*, ed. A. Egziabher et al. Ottawa: IDRC.

Lefebvre, H. 1991. *The production of space.* Malden, MA: Blackwell.

Lefebvre, H. 1996 [1968]. *Writings on cities.* Oxford: Blackwell.

Loram, A., J. Tratalos, P.H. Warren, and K.J. Gaston. 2007. Urban domestic gardens (X): The extent & structure of the resource in five major cities. *Landscape Ecology* 22 (4): 601–615.

Martellozzo, F., J.S. Landry, D. Plouffe, V. Seufert, P. Rowhani, and N. Ramankutty. 2014. Urban agriculture: A global analysis of the space constraint to meet urban vegetable demand. *Environmental Research Letters* 9 (6): 1–9.

Masvaure, S. 2016. Coping with food poverty in cities: The case of urban agriculture in Glen Norah Township in Harare. *Renewable Agriculture and Food Systems* 31 (3): 202–213.

Maxwell, D.G. 1994. The household logic of urban farming in Kampala. In *Cities feeding people: An examination of urban agriculture in East Africa*, ed.

A.G. Egziabher, D. Lee-Smith, D.G. Maxwell, P.A. Memon, L.J.A. Mougeot, and C.J. Sawio. Ottawa: IDRC.

Maxwell, D., C. Levin, M. Armar-Klemesu, M. Ruel, and S. Morris. 2000. *Urban livelihoods and food and nutrition security in greater Accra, Ghana.* Washington, DC, US: IFPRI.

Mayer, P. (1971). *Townsmen Or tribesmen: Conservatism and the process of urbanization in a South African city.* Oxford University Press.

Mbaye, A., and P. Moustier. 2000. Market-oriented urban agricultural production in Dakar. *Growing cities, growing food: Urban agriculture on the policy agenda*, 235–256.

Mbiba, B. 1995. *Urban agriculture in Zimbabwe: Implications for urban management and poverty.* Aldershot: Avebury.

Mbiba, B. 2000. Urban agriculture in Harare: Between suspicion and repression. In *Growing cities, growing food: Urban agriculture on the policy agenda—A reader on urban agriculture*, ed. N. Bakker, M. Dubbeling, S. Gündel, U. Sabel-Koschella, and H. de Zeeuw, 285–301. Feldafing: GTZ.

Mbiba, B. 2001. The marginalisation of urban agriculture in Lusaka. *Urban Agriculture Magazine* 4: 19–21.

Medearis, D., and W. Daseking. 2012. Freiburg, Germany: Germany's eco-capital. In *Green cities of Europe*, 65–82. Washington, DC: Island Press.

Melbourne Planning Authority. 2008. Creating liveable new communities—Promising practice: A book of 'good practice' case studies. http://www.mpa.vic.gov.au/wp-content/Assets/Files/Promising%20Practice%209%20April%2008.pdf. Accessed 1 Feb 2016.

Milan Urban Food Policy Pact. 2015. http://www.milanurbanfoodpolicypact.org.

Mitlin, D., and J. Mogaladi. 2013. Social movements and the struggle for shelter: A case study of eThekwini (Durban). *Progress in Planning* 84: 1–39.

Mkwela, H.S. 2014. Governance of agriculture in the cities of developing countries: Local leaders' perspectives. *Journal of Geography and Regional Planning* 7 (3): 36.

Mlozi, M.R. 1996. Urban agriculture in Dar es Salaam: Its contribution to solving the economic crisis and the damage it does to the environment. *Development Southern Africa* 13 (1): 47–65.

Mougeot, L.J.A. 1994. *Urban food production: Evolution, official support and significance.* Cities feeding people report 8. Ottawa: IDRC.

Mtani, A. 1997. Urban agriculture in Dar es Salaam. *Geographical Journal of Zimbabwe* 28: 49–58.

Mubvami, T., S. Mushamba, and H. De Zeeuw. 2006. Integration of agriculture in urban land use planning. *Cities Farming for the Future: Urban Agriculture for Green and Productive Cities.* RUAF, IIRR and IDRC, Silang, the Philippines, 54–74.

Mujere, N. 2017. The Contribution of Smallholder Irrigated Urban Agriculture Towards Household Food Security in Harare, Zimbabwe. Ch 18. *Global Urban Agriculture*, 220.

National Farmer Federation. 2017. http://www.nff.org.au/farm-facts.html.

Novo, M.G., and C. Murphy. 2000. Urban agriculture in the city of Havana: A popular response to a crisis. In *Growing cities, growing food. Urban agriculture on the policy agenda*, ed. N. Bakker, M. Dubbeling, S. Gündel, U. Sabel-Koshella, and H. de Zeeuw, 329–346. Feldafing, Germany: Zentralstelle für Ernährung und Landwirtschaft (ZEL).

Nunan, F. 2000. Waste recycling through urban farming in Hubli-Dharwad. In *crowing gities, frowing food: Urban agriculture on the policy agenda*, ed. N. Bakker, M. Dubbeling, S. Guendel, U. Sabel-Koaschella, and H. Zeeuw de. Eurasburg: DSE.

Opitz, I., R. Berges, A. Piorr, and T. Krikser. 2016. Contributing to food security in urban areas: Differences between urban agriculture and peri-urban agriculture in the Global North. *Agriculture and Human Values* 33 (2): 341–358.

Pires, V. 2011. Planning for urban agriculture planning in Australian cities. State of Australian Cities.

Popovitch, T. 2014. 10 American cities lead the way with urban agriculture ordinances. Available at: http://seedstock.com/2014/05/27/10-american-cities-lead-the-way-with-urban-agriculture-ordinances/. Accessed 29 May 2014.

Potutan, G.E., W.H. Schnitzler, J.M. Arnado, L.G. Janubas, and R.J. Holmer. (2000). Urban agriculture in Cagayan de Oro: A favourable response of city government and NGOs. *Growing cities, growing food: urban agriculture on the policy agenda*, 413–428.

Potter, R., and T. Unwin (eds.). 1989. *The geography of urban-rural interaction in developing countries*. London: Routledge.

Prain, G., N. Gonzales, B. Arce, and J. Tenorio. 2010. Organic vegetable production on the peri-urban interface: Helping low income producers access high value markets in Lima, Peru. *Acta Hort (ISHS)* 881: 117–123.

Purnomohadi, N. 2000. Jakarta: Urban agriculture as an alternative strategy to face the economic crisis. In *Growing cities, growing food. Urban agriculture on the policy agenda*, ed. N. Bakker, M. Dubbeling, S. Gündel, U. Sabel-Koshella, and H. de Zeeuw, 453–465. Feldafing, Germany: Zentralstelle für Ernährung und Landwirtschaft (ZEL).

Sabiiti, E.N., and C.B. Katongole. 2014. Urban agriculture: A response to the food supply crisis in Kampala city, Uganda. In *The security of water, food, energy and liveability of cities*, ed. B. Maheshwari, R. Purohit, H. Malano, V. Singh, and P. Amerasinghe, vol. 71. Dordrecht: Water Science and Technology Library, Springer.

Sanyal, B. 1987. Urban cultivation amidst modernisation: How should we interpret it? *Journal of Planning Education and Research* 6: 187–207.

Sawio, C. 1994. Who are the farmers of Dar es Salaam? In *Cities feeding people: An examination of urban agriculture in East Africa*, ed. A.G. Egziabher et al., 23–44. Ottawa: International Development Research Centre.

Smith, D., and D. Tevera. 1997. Socio-economic context for the householder of urban agriculture in Harare, Zimbabwe. *Geographical Journal of Zimbabwe* 28: 25–38.

Soja, E. 2010. Spatialising the urban, part 1. *City: Analysis of Urban Trends, Culture, Theory, Policy, Action* 14 (6): 629–635.

Tacoli, C. 1998. Rural-urban interactions: A guide to the literature. *Environment and Urbanization* 10 (1): 147–166.

Thornton, A. 2008. Beyond the metropolis: Small town case studies of urban and peri-Urban agriculture in South Africa. *Urban Forum* 19 (3): 243-262.

Thornton, A. 2012. *Urban agriculture in South Africa: A study of the Eastern Cape province*. Lewiston, NY: Edwin Mellen Press.

Thornton, A. 2017. "The Lucky country"? A critical exploration of community gardens and city–community relations in Australian cities. *Local Environment* 22 (8): 969–985.

Torres-Lima, P., L.M. Rodriques Sanchez, and B.I. Garcia Uriza. 2000. Mexico City: The integration of urban agriculture to contain urban sprawl. In *Growing cities, growing food. Urban agriculture on the policy agenda*, ed. N. Bakker, 363–390. Germany: DSE.

Von Thünen, J.H. 1875. In *Der isolirte staat in beziehung auf landwirtschaft und nationalökonomie*, vol. 1. Wiegant, Hempel & Parey.

White, C., K. Woodfield, J. Ritchie, and R. Ormston. 2014. Writing up qualitative research. In *Qualitative research practice: A guide for social science students and researchers*, ed. J. Ritchie, J. Lewis, C.M. Nicholls, and R. Ormston, 367–400. London: Sage.

Yang, Z., J. Cai, and R. Sliuzas. 2010. Agro-tourism enterprises as a form of multi-functional urban agriculture for peri-urban development in China. *Habitat International* 34 (4): 374–385.

Yi-Zhang, C., and Z. Zhangen. 2000. Shanghai: Trends towards specialised and capital-intensive urban agriculture. *Growing cities growing food: Urban agriculture on the policy agenda*, 467–477. Feldafing: DSE-ZEL.

Yoveva, A., B. Gocheva, G. Voykova, B. Borissov, and A. Spassov. 2000. Sofia: Urban agriculture in an economy in transition. *Growing cities, growing food: Urban agriculture on the policy agenda. A reader on urban agriculture*, 501–518.

CHAPTER 4

Conclusion

Abstract This book discussed some of the research in this space from cities in the 'North and South'. Exploring examples from the past of 'War' and 'Victory Gardens', where domestic cross-sectoral and multi-actor allegiances formed to address an international or global crisis, provides a bridge to current urban social processes for city-community engagement in food systems change. In the former, world war provided the context for national, local and grassroots partnerships and, for the latter, the crisis is unpredictable in the global food system due to oil price shocks and climate change. Urban activism in North cities are engaged in hard fought campaigns for the preservation of urban space for non-economic uses, such as urban food production, and other social uses. The emergence of urban food activism in South cities, particularly is SSA, is impaired by poor urban governance that does not promote and encourage city-community partnerships in the use of space for social use value.

Keywords Critical urban theory · Community gardens
Food Policy Councils · Urban agriculture · Neoliberal · Social
production Socio-spatial · Urban geographies · Food supply chain

Critical urban theorists have questioned the organisation of urban space, its power structures, conditions and inequality (Mitchell 2003; Harvey 2008, 2012; Soja 2010). Conceptualising urban agriculture as an alternative social production of space is an emerging area for urban theorists

© The Author(s) 2018   101
A. Thornton, *Space and Food in the City*,
https://doi.org/10.1007/978-3-319-89324-2_4

and critical geographers. Eizenberg (2012) explores community gardens in New York City and its sustainability as an alternative social space in the neoliberal city. Crane et al. (2013) explored community gardens and 'guerrilla gardening' in Ontario Canada, to assess if the latter is a particular 'type' of urban agriculture that is distinct from community gardening, as unique form of social resistance. These studies, although unique in applying critical urban geographies to explore particular aspects of urban agriculture, do not, however, consider the implications of its appropriation by the city, as a larger citywide project, for sustainable urban development. In cities in developed and developing countries, citizens are seeking change in 'the locus of lived experience', whereby they are seeking social transformation through changing space (Lefebvre 1991).

With various social and food justice themes increasingly playing out in cities, globally, these issues are ideally suited for a critical urban lens. This book explored the expanding literature on critical urban theory that provided a useful lens for case study research on urban activism for urban food systems change. Whereas critical urban theorists point to reproduction of urban spaces as a reflection of urban dwellers asserting their views and needs in space, urban agriculture can be conceptualised as the social and physical embodiment of this 'right' in creating 'fully lived' and more 'just' urban landscapes. In the wider literature, it appears that local councils in western cities are willing to respond to demands for locally based urban and peri-urban food production and marketing at a point when community and market gardens activities reach a critical mass, alongside consumer support and broader advocacy for these activities. Once integrated into formal urban planning spaces, it is not quite clear to what extent that these local and alternative (as opposed to mainstream) food system activities continue to satisfy needs for social, food and environmental justice.

The discussion of Food Policy Councils (FPC) offers an example of integrated community-led urban food systems. Originating nearly 40 years ago (in Knox, Tennessee), many western cities has adopted local FPCs to achieve these benefits, as part of an integrated city-community planning approach to developing an urban 'food-shed'. The creation of FPCs is relevant, as it can provide an arena for discussion and strategy development for a socially and environmentally resilient urban food system, through engaging with a range of stakeholders from the following food system sectors: production, processing, distribution, consumption and waste recycling.

There appears to be an attempt by some city councils to work along-side community groups in low-income neighbourhoods to facilitate local food and marketing systems for local economic and community development. As McClintock (2014) observed, contradictions between neoliberalism and alternative food systems, such as urban agriculture, need to be understood as both existing within the capitalist market logic and as a public good. To a certain extent, Boston city planners are attempting to integrate economic and social use values through commercial zoning and establishing the legality of urban agriculture and local food networks, and as being particularly beneficial for low-income neighbourhoods. Further study might explore to what extent these initiatives are a form of urban 'green-washing', or sincere attempts at the policy level to include grass-roots voices in decision-making regarding the social use of urban space.

Discussed in earlier chapters, the roots of an urban (and peri-urban) food growing culture in global North cities can be traced back to its role in strengthening domestic food supplies during times of economic crisis and global conflict. The concept of urban and peri-urban food production to achieve desired outcomes, as determined by citizen and state (city/municipal) actors, is not unusual to cities in developed countries. This interplay between 'city and citizens' in creating urban socio-spatial policymaking is critical to urban governance. Knowledgeable urban dwellers, as urban activists, mobilise public and private resources to push grassroots agendas, such as urban food system change, on to their elected officials. For cities in the South, the notion of urban food activism is intriguing, as this form of social mobilisation could challenge weak urban governance structures and move UA from its status as a fringe activity. Until a culture of urban activism becomes a reality, UA will remain as a fringe activity, and unpopular in the neoliberal city, where there is little appetite for non-market, community-led solutions to the socio-spatial failings of the neoliberal dominated urban food supply chain.

REFERENCES

Crane, A., L. Viswanathan, and G. Whitelaw. 2013. Sustainability through intervention: A case study of guerrilla gardening in Kingston, Ontario. *Local Environment: The International Journal of Justice and Sustainability* 18 (1): 71–90.

Eizenberg, E. 2012. Actually existing commons: Three moments of space in community gardens in New York City. *Antipode* 44 (3): 764–782.

Harvey, D. 2008. The right to the city. *The New Left Review* 53 (Sept–Oct): 23–40.

Harvey, D. 2012. *Rebel cities: From the right to the city to the urban revolution*. London: Verso.

Lefebvre, H. 1991. *The production of space*. Malden, MA: Blackwell.

McClintock, N. 2014. Radical, reformist, and garden-variety neoliberal: Coming to terms with urban agriculture's contradictions. *Local Environment: The International Journal of Justice and Sustainability* 19 (2): 147–171.

Mitchell, D. 2003. *The right to the city: Social justice and the fight for public space*. Guilford Press.

Soja, E. 2010. Spatialising the urban. Part 1. *City: Analysis of Urban Trends, Culture, Theory, Policy, Action* 14 (6): 629–635.

REFERENCES

Acton, L. 2011. Allotment gardens: A reflection of history, heritage, community and self. *Papers from the Institute of Archaeology* 21: 46.

Adger, W.N. 2000. Social and ecological resilience: Are they related? *Progress in Human Geography* 24 (3): 347–364.

Adie, K. 2013. *Fighting on the homefront: The legacy of women in World War One*. London: Hodder & Stoughton.

Alkon, A.H. 2013. Food justice, food sovereignty and the challenge of neoliberalism. In *Food sovereignty: A critical dialogue*, September, 13–14.

American Planning Association [APA]. 2010. Zoning for urban agriculture. In *Zoning for Practice March 2010*. Available at: https://www.planning.org/publications/publication/9006942/. Accessed 16 Apr 2018.

Armar-Klemesu, M., and D. Maxwell. 2000. Accra: Urban agriculture as an asset strategy, supplementing income and diets. *Growing cities, growing food: Urban agriculture on the policy agenda*, 203–208.

Australian City Farms and Community Gardens Network. 2011. https://communitygarden.org.au/2011/12/21/sydney-city-community-gardens-attract-visitors/.

Australian Government. 2013. The state of Australian cities. https://infrastructure.gov.au/infrastructure/pab/soac/index.aspx. Accessed 22 Jan 2016.

Badami, M.G., and N. Ramankutty. 2015. Urban agriculture and food security: A critique based on an assessment of urban land constraints. *Global Food Security* 4: 8–15.

Barbour, R.S. 2001. Checklists for improving rigour in qualitative research: A case of the tail wagging the dog? *BMJ: British Medical Journal* 322 (7294): 1115–1117.

Barnett, C., P. Cloke, N. Clarke, and A. Malpass. 2005. Consuming ethics: Articulating the subjects and spaces of ethical consumption. *Antipode* 37: 23–45.

Barthel, S., J. Parker, and H. Ernstson. 2015. Food and green space in cities: A resilience lens on gardens and urban environmental movements. *Urban Studies* 52 (7): 1321–1338.

Battersby, J. 2011. Urban food insecurity in Cape Town, South Africa: An alternative approach to food access. *Development Southern Africa* 28 (4): 545–561.

Battersby, J. 2012. Beyond the food desert: Finding ways to speak about urban food security in South Africa. *Geografiska Annaler: Series B, Human Geography* 94 (2): 141–159.

Beall, J. 2002. Globalization and social exclusion in cities: Framing the debate with lessons from Africa and Asia. *Environment and Urbanization* 14 (1): 41–51.

Beall, J., O. Crankshaw, and S. Parnell. 2014. *Uniting a divided city: Governance and social exclusion in Johannesburg.* London: Routledge.

Beck, U. 1999. *World risk society.* Cambridge, UK: Polity Press.

Bek, D., T. Binns, and E. Nel. 2004. 'Catching the development train': Perspectives on 'top-down' and 'bottom-up' development in post-apartheid South Africa. *Progress in Development Studies* 4 (1): 22–46.

Bengtsson, M. 2016. How to plan and perform a qualitative study using content analysis. *NursingPlus Open* 2: 8–14.

Bohstedt, J. 2014. Food riots and the politics of provisions in world history. *IDS Working Papers* 2014 (444): 1–31.

Borel-Saladin, J. 2017. Where to draw the line: Data problems and other difficulties estimating urbanisation in Africa. *Development Southern Africa* 34 (2): 137–150.

Bourne, G. 1942. Feeding post-war Europe. *Nature* 149: 182–184.

Brenner, N. 2000. The urban question as a scale question: Reflections on Henri Lefebvre, urban theory and the politics of scale. *International Journal of Urban and Regional Research* 24 (2): 361–378.

Brenner, N. 2012. What is critical urban theory? In *Cities for people, not for profit: Critical urban theory and the right to the city,* ed. N. Brenner, P. Marcuse, and M. Mayer, 11–23. New York: Routledge.

Brenner, N., and S. Elden. 2001. Henri Lefebvre in contexts: An introduction. *Antipode* 33 (5): 763–768.

Brenner, N., and N. Theodore. 2002. Cities and the geographies of "actually existing neoliberalism". *Antipode* 34 (3): 349–379.

Brown, A. 2013. The right to the city: From Paris 1968 to Rio 2010. *International Journal of Urban and Regional Research* 37 (3): 957–971.

Brown, K., M. Bailkey, A. Meares-Cohen, J. Nasr, J. Smit, T. Buchanan, and P. Mann. 2002. Urban agriculture and community food security in the United States: Farming from the city center to the urban fringe. Available at http://www.foodsecurity.org/PrimerCFSCUAC.pdf. Accessed 20 Dec 2017.

Bruntland, G.H. 1987. Our common future: Report of the World Commission on Environment and Development. World Commission on Environment and Development.

Bundy, C. 1979. *The rise and fall of the South African peasantry.* London: Heinemann.

Bush, R. 2010. Food riots: Poverty, power and protest. *Journal of Agrarian Change* 10 (1): 119–129.

Castells, M. 1983. *The city and the grassroots: A cross-cultural theory of urban social movements* (no. 7). Berkeley: University of California Press.

Cobbinah, P., M. Erdiaw-Kwasie, and P. Amoateng. 2015. Africa's urbanisation: Implications for sustainable development. *Cities* 47: 62–72.

Colding, J., and S. Barthel. 2013. The potential of 'urban green commons' in the resilience building of cities. *Ecological Economics* 86: 156–166.

Companioni, N., Y. Hernandez, E. Paez, and C. Murphy. 2002. The growth of urban agriculture. In *Sustainable agriculture and resistance: Transforming food production in Cuba*, ed. F. Funes, L. Garcia, M. Bourque, and P. Rosset. Oakland, CA: Food First.

Conservation Law Foundation. 2012. Growing green: Measuring benefits, overcoming barriers, and nurturing opportunities for urban agriculture in Boston. http://clf.org/growing-green/. Accessed 3 Feb 2014.

Corden, A., and R. Sainsbury. 2006. *Using verbatim quotations in reporting qualitative social research: Researchers' views*, 11–14. York, UK: University of York.

Crane, A., L. Viswanathan, and G. Whitelaw. 2013. Sustainability through intervention: A case study of guerrilla gardening in Kingston, Ontario. *Local Environment: The International Journal of Justice and Sustainability* 18 (1): 71–90.

Crush, J., and B. Frayne. 2011. Urban food insecurity and the new international food security agenda. *Development Southern Africa* 28 (4): 527–544.

Crush, J., A. Hovorka, and D. Tevera. 2011. Food security in Southern African cities: The place of urban agriculture. *Progress in Development Studies* 11 (4): 285–305.

Dasso, A., and T. Pinzas. 2000. NGO experiences in Lima targeting the urban poor through urban agriculture. In *Growing cities, growing food-urban agriculture on the policy agenda*, ed. N. Bakker, M. Dubbeling, S. Guendel, U. Sabel-Koschella, and H. de Zeeuw, 349–362. Eurasburg, Germany: DSE.

De Haen, H. 2002. Enhancing the contribution of urban agriculture to food security. Urban Agriculture Magazine. Special issue, *For the world food summit: Five years later*. The Netherlands, Resource Centre for Urban Agriculture (RUAF).

Della Porta, D., and M. Diani. 2006. *Social movements: An introduction*, 2nd ed. Oxford: Blackwell.

Desai, M. 2014. *The paradigm of international social development: Ideologies, development systems and policy approaches*. New York: Routledge.

Detroit Free Press. 2011. *Bill would create right to farm act exemption for Detroit* [online]. Available at http://www.freep.com/article/20111128/ NEWS06/111280346/Bill-would-create-Right-Farm-Act-exemption-Detroit. Accessed 14 Mar 2014.

Dieleman, H. 2016. Urban agriculture in Mexico City; balancing between ecological, economic, social and symbolic value. *Journal of Cleaner Production*. https://doi.org/10.1016/j.jclepro.2016.01.082.

Donald, B., and A. Blay-Palmer. 2006. The urban creative-food economy: Producing food for the urban elite or social inclusion opportunity? *Environment and Planning A* 38: 1901–1920.

Downe-Wambolt, D. 1992. Content analysis: Method, applications and issues. *Health Care for Women International* 13: 313–321.

Drescher, A.W. 2001. The German allotment gardens: A model for poverty alleviation and food security in Southern African cities. In *Proceedings of the sub-regional expert meeting on urban horticulture, Stellenbosch, South Africa*, January, 159–167.

Dryzek, J.S., D. Downes, C. Hunold, D. Schlosberg, and H.K. Hernes. 2003. *Green states and social movements: Environmentalism in the United States, United Kingdom, Germany, and Norway*. Oxford: OUP.

DuPuis, E.M., and D. Goodman. 2005. Should we go "home" to eat?: Toward a reflexive politics of localism. *Journal of Rural Studies* 21 (3): 359–371.

Eckhardt, F., and Ingemar Elander. 2009. Urban governance: Introduction. In *Urban governance in Europe*, ed. F. Eckhardt and Ingemar Elander. Berlin: BWV Berliner Wissenschafts-Verlag.

Egziabher, A. 1994. Urban farming, co-operatives and the urban poor in Addis Ababa. In *Cities feeding people: An examination of urban agriculture in East Africa*, ed. A. Egziabher, 67–84. Ottawa: International Development Research Centre.

Eizenberg, E. 2012. Actually existing commons: Three moments of space in community gardens in New York City. *Antipode* 44 (3): 764–782.

Elden, S. 2004. Between Marx and Heidegger: Politics, philosophy and Lefebvre's 'the production of space'. *Antipode* 36 (1): 86–105.

Ellis, F., and J. Sumberg. 1998. Food production, urban areas and policy responses. *World Development* 26 (2): 213–225.

Endres, A.B., and J.M. Endres. 2009. Homeland security planning: What victory gardens and Fidel Castro can teach us in preparing for food crises in the United States. *Food and Drug Law Journal* 64: 405.

Fan, S., L. Zhang, and X. Zhang. 2004. Reforms, investment, and poverty in rural China. *Economic Development and Cultural Change* 52 (2): 395–421.

Feagan, R. 2007. The place of food: Mapping out the 'local' in local food systems. *Progress in Human Geography* 31 (1): 23–42.

Fisher, D.R., and W.R. Freudenburg. 2001. Ecological modernization and its critics: Assessing the past and looking toward the future. *Society and Natural Resources* 14: 701–709.

Foeken, D., M. Sofer, and M. Mlozi. 2004. *Urban agriculture in Tanzania: Issues of sustainability.* African Studies Centre research report 75. Leiden.

Folke, C. 2006. Resilience: The emergence of a perspective for social–ecological systems analyses. *Global Environmental Change* 16: 253–267.

Foo, K., D. Martin, C. Wool, and C. Polsky. 2014. Reprint of "The production of urban vacant land: Relational placemaking in Boston, MA neighborhoods". *Cities* 40: 175–182.

Food and Agricultural Organisation (FAO) of the United Nations. 2012. *Growing greener cities in Africa.* First status report on urban and peri-urban horticulture in Africa. Rome: FAO.

Food and Agricultural Organisation (FAO). 2013. *The state of food insecurity in the world 2013: The multiple dimensions of food security.* Rome: Food and Agriculture Organization of the United Nations.

Food and Agricultural Organisation (FAO) of the United Nations. 2015. *The state of food insecurity in the world (SOFI).* Rome: Food and Agriculture Organization (FAO) of the United Nations.

Foster, J.B. 2002. *Ecology against capitalism.* New York: Monthly Review Press.

Frayne, B., W. Pendleton, J. Crush, B. Acquah, J. Battersby-Lennard, E. Bras, et al. 2010. *The state of urban food insecurity in southern Africa.* Urban Food Security Series, 2. Kingston and Cape Town: Queen's University and AFSUN.

Freeman, D.B. 1991. *City of farmers: Informal urban agriculture in the open spaces of Nairobi, Kenya.* McGill-Queen's Press-MQUP.

Freestone, R. 2014. Australian environmental planning, origin and theories. In *Australian environmental planning: Challenges and future prospects,* ed. J. Byrne, N. Sipe, and J. Dodson, 21–35. New York: Routledge.

Garnett, T. 2001. Urban agriculture in London: Rethinking our food economy. In *Growing cities, growing food: Urban agriculture on the policy agenda,* ed. N. Bakker, M. Dubbeling, S. Gündel, U. Sabel-Koschella, and H. De Zeeuw, 477–500. Feldafing: DSE/Zentralstelle für Ernährung und Landwirtschaft.

Gaynor, A. 2006. *Harvest of the suburbs. An environmental history of growing food in Australian cities.* Crawley: University of Western Australia Press.

Gertel, J., and S. Samir. 2000. Cairo: Urban agriculture and visions for a 'modern' city. In *Growing cities, growing food: Urban agriculture on the policy agenda,* 209–234. Feldafing: Deutsche Stiftung für Internationale Entwicklung (DSE), Zentralstelle für Ernährung und Landwirtschaft.

Ghosh, J. 2010. The unnatural coupling: Food and global finance. *Journal of Agrarian Change* 10 (1): 72–86.

Gibson-Graham, J.K. 2008. Diverse economies: Performative practices for 'other worlds'. *Progress in Human Geography* 32 (5): 613–632.

Gogwana, M. 2001. The role of urban agriculture in food security: A case of low-income dwellers in Dangamvura. In 2nd WARFSA/WaterNet Symposium: Integrated Water Resource Management: Theories, Practices, Cases, 30–31 October, Cape Town.

Goodman, D. 2003. The quality 'turn' and alternative food practices: Reflections and agenda. *Journal of Rural Studies* 19: 1–7.

Goodman, D., M. DuPuis, and M. Goodman. 2011. *Alternative food networks: Knowledge, practice and politics.* London: Routledge.

Gröning, G. 1996. Politics of community gardening in Germany. Available at http://www.cityfarmer.org/german99.html. Accessed 27 June 2017.

Gurran, N. 2011. *Australian urban land use planning: Principles, systems and practice.* Sydney: Sydney University Press.

Guthman, J. 2008. Neoliberalism and the making of food politics in California. *Geoforum* 39 (3): 1171–1183.

Hajer, M.A. 1995. *The politics of environmental discourse: Ecological modernization and the policy process,* 40. Oxford: Clarendon Press.

Hallett, S., L. Hoagland, E. Toner, T.M. Gradziel, C.A. Mitchell, and A.L. Whipkey. 2016. Urban agriculture: Environmental, economic, and social perspectives. *Horticultural Reviews* 44: 65–120.

Hamilton, A.J., K. Burry, H.F. Mok, S.F. Barker, J.R. Grove, and V.G. Williamson. 2014. Give peas a chance? Urban agriculture in developing countries. A review. *Agronomy for Sustainable Development* 34 (1): 45–73.

Hammond, D. 2003. *The science of synthesis: Exploring the social implications of general systems theory.* Boulder: University Press of Colorado.

Harvey, D. 2003. *The new imperialism.* Oxford: Oxford University Press.

Harvey, D. 2008. The right to the city. *The New Left Review* 53 (Sept–Oct): 23–40.

Harvey, D. 2012. *Rebel cities: From the right to the city to the urban revolution.* London: Verso.

Hayden-Smith, R. 2014. *Sowing the seeds of victory: American gardening programs of World War 1.* Jefferson, NC: McFarland.

Healey, P. 2004. Creativity and urban governance. *disP-The Planning Review* 40 (158): 11–20.

Heimlich, R., and C. Bernard. 1993. *Agricultural adaptation to urbanization: Farm types in the United States metropolitan area.* Washington, DC: USDA, Economic Research Service.

Henderson, B.R., and K. Hartsfield. 2009. Is getting into the community garden business a good way to engage citizens in local government? *National Civic Review* 98 (4): 12–17.

Hodgson, K., M.C. Campbell, and M. Bailkey. 2011. *Urban agriculture: Growing healthy, sustainable places.* Washington, DC: APA Planning Advisory Service.

Hopkins, R. 2008. *The transition handbook: From oil dependency to local resilience.* Chelsea: Green Publishing.

Hossain, N. 2009. *Towards a new social justice agenda: Understanding political responses to crises.* IDS InFocus Policy Briefing, issue 11 October 2009. Brighton: Institute for Development Studies.

Huchzermeyer, M. 2012. Informal settlements: Production and intervention in Brazil and South Africa. *Latin American Perspectives* 29 (1): 83–105. New York: Sage.

Huchzermeyer, M. 2014. Humanism, creativity and rights: Invoking Henri Lefebvre's right to the city in the tension presented by informal settlements in South Africa today. *Transformation: Critical Perspectives on Southern Africa* 85 (1): 64–89.

Ibrahim, A.A., and S.A. Elariane. 2018. Feasibility tools for urban animal husbandry in cities: Case of Greater Cairo. *Urban Research & Practice* 11 (2): 111–138.

International Assessment of Agricultural Science, Technology and Development (IAASTD). 2009. *Agriculture at a crossroads: Global report.* Washington, DC: Island Press.

Islam, M., and N.C. Nag. 2010. *Economic integration in South Asia: Issues and pathways.* Delhi: Pearson Education India.

Jacobi, P., J. Amend, and S. Kiango. 2000. Urban agriculture in Dar es Salaam: Providing an indispensable part of the diet. *Growing cities, growing food: Urban agriculture on the policy agenda*, 257–283.

Jarosz, L. 2008. The city in the country: Growing alternative food networks in Metropolitan areas. *Journal of Rural Studies* 24 (3): 231–244.

Jarosz, L. 2009. The political economy of global governance and the world food crisis: The case of the FAO. *Review* 332 (1): 37–60.

Johannesburg Development Agency. 2015. Rooftop garden plants seeds of success. Available at: http://www.jda.org.za/index.php/corporate-social-responsibility/1542-rooftop-garden-plants-seeds-of-success2. Accessed 20 Oct 2017.

Kaufman, M., and J. Read. 2016. A case study: Legalizing commercial agriculture in Boston—A logical step towards integrating farming into urban life. In *Sowing seeds in the city*, ed. S. Brown, K. McIvor, and E. Hodges Snyder. Dordrecht: Springer.

Kneafsey, M. 2010. The region in food—Important or irrelevant? *Cambridge Journal of Regions, Economy and Society* 3 (2): 177–190.

Kreinecker, P. 2000. La Paz: Urban Agriculture in harsh ecological conditions. In *Growing cities, growing food*, ed. N. Bakker, M. Dubbeling, S. Güendel, U. Sabel-Koschella, and H. de Zeeuw, 391–412. Feldafing: DSE.

Krishnan, S., D. Nandwani, G. Smith, and V. Kankarta. 2016. Sustainable urban agriculture: A growing solution to urban food deserts. In *Organic farming for sustainable agriculture*, 325–340. Cham: Springer.

Kühn, M. 2003. Greenbelt and green heart: Separating and integrating landscapes in European city regions. *Landscape Urban Planning* 64 (1–2): 19–27. https://doi.org//10.1016/S0169-2046(02)00198-6.

Land Stewardship Project. 2010. How U.S. cities are using zoning to support urban agriculture. *LSP Factsheet* #21, 1–2. Available at www.landstewardship-project.org/repository/1/253/urbanagzoning.pdf. Accessed 24 Jan 2016.

Larder, N., K. Lyons, and G. Woolcock. 2012. Enacting food sovereignty: Values and meanings in the act of domestic food production in urban Australia. *Local Environment: The International Journal of Justice and Sustainability* 19 (1): 56–76. https://doi.org//10.1080/13549839.2012.716409.

Larder, N., K. Lyons, and G. Woolcock. 2014. Enacting food sovereignty: Values and meanings in the act of domestic food production in urban Australia. *Local Environment* 19 (1): 56–76.

Lawson, L.J. 2005. *City bountiful. A century of community gardening in America*. Berkeley, Los Angeles, CA, and London, UK: University of California Press.

Lawson, L.J. 2014. Garden for victory! The American victory garden campaign of World War II. In *Greening in the Red Zone: Disaster, resilience and community greening*, ed. K.G. Tidball and M. Krasny. Dordrecht: Springer.

Ledoux, T.F., and I. Vojnovic. 2013. Going outside the neighborhood: The shopping patterns and adaptations of disadvantaged consumers living in the lower eastside neighborhoods of Detroit, Michigan. *Health & Place* 19: 1–14.

Lee-Smith, D., and P. Ali Memon. 1994. Urban agriculture in Kenya. In *Cities feeding people: An examination of urban agriculture in East Africa*, ed. A. Egziabher et al. Ottawa: IDRC.

Lefebvre, H. 1991. *The production of space*. Malden, MA: Blackwell.

Lefebvre, H. 1996 [1968]. *Writings on cities*. Oxford: Blackwell.

Lefebvre, H. 2003 [1970]. *The urban revolution*. Minneapolis: University of Minnesota Press.

Lehmann, S. 2010. *Principles of green urbanism: Transforming the city for sustainability*. London: Earthscan.

Lever, J., and A. Evans. 2017. Corporate social responsibility and farm animal welfare: Towards sustainable development in the food industry? In *Stages of corporate social responsibility*, 205–222. Cham: Springer.

Leyshon, A., R. Lee, and C. Williams (eds.). 2003. *Alternative economic spaces*. London: Sage.

Lindell, I. 2008. The multiple sites of urban governance: Insights from an African city. *Urban Studies* 45 (9): 1879–1901.

Loram, A., J. Tratalos, P.H. Warren, and K.J. Gaston. 2007. Urban domestic gardens (X): The extent & structure of the resource in five major cities. *Landscape Ecology* 22 (4): 601–615.

Lourenco-Lindell, I. 1997. Food for the poor, food for the city: The role of urban agriculture in Bissau. *Geographical Journal of Zimbabwe* (28): 39–48.

Low, S.A., and S. Vogel. 2011. *Direct and intermediated marketing of local foods.* United States Department of Agriculture, Economic Research Report No. 128. U.S. Department of Agriculture, Economic Research Service.

Lynch, K., R. Maconachie, T. Binns, P. Tengbe, and K.S. Bangura. 2012. Meeting the urban challenge? Urban agriculture and food security in post-conflict Freetown, Sierra Leone. *Applied Geography* 36 (1): 31–39.

Lyons, K., C. Richards, L. DesFours, and M. Amati. 2013. Food in the city: Urban food movements and the (re)imagining of urban spaces. *Australian Planner* 50 (2): 157–163.

Lyson, T.A. 2004. *Civic agriculture: Reconnecting farm, food and community.* Metford: Tufts University Press.

Mackay, H. 2018. Mapping and characterising the urban agricultural landscape of two intermediate-sized Ghanaian cities. *Land Use Policy* 70: 182–197.

Mah, C.L., and H. Thang. 2013. Cultivating food connections: The Toronto food strategy and municipal deliberation on food. *International Planning Studies* 18 (1): 96–110.

Malan, N. 2015. Urban farmers and urban agriculture in Johannesburg: Responding to the food resilience strategy. *Agrekon* 54 (2): 51–75.

Marcuse, P. 2009. From critical urban theory to the right to the city. *City* 13 (2–3): 185–197.

Marsden, T. 2000. Food matters and the matter of food: Towards a new food governance? *Sociologia Ruralis* 40 (1): 20–29.

Marsden, T., J. Banks, and G. Bristow. 2000. Food supply chain approaches: Exploring their role in rural development. *Sociologia Ruralis* 40: 424–438.

Martellozzo, F., J.S. Landry, D. Plouffe, V. Seufert, P. Rowhani, and N. Ramankutty. 2014. Urban agriculture: A global analysis of the space constraint to meet urban vegetable demand. *Environmental Research Letters* 9 (6): 1–9.

Martin, G. 2011. Going local: Quantifying the economic impacts of buying from locally owned businesses in Portland, Maine. Maine Center for Economic Policy.

Martinez, S., M. Hand, M. Da Pra, S. Pollack, K. Ralston, T. Smith, ... and C. Newman. 2010. *Local food systems: Concepts, impacts, and issues* (No. 24313), University Library of Munich, Germany.

Masvaure, S. 2016. Coping with food poverty in cities: The case of urban agriculture in Glen Norah Township in Harare. *Renewable Agriculture and Food Systems* 31 (3): 202–213.

Maxwell, D.G. 1994. The household logic of urban farming in Kampala. In *Cities feeding people: An examination of urban agriculture in East Africa*, ed. A.G. Egziabher, D. Lee-Smith, D.G. Maxwell, P.A. Memon, L.J.A. Mougeot, and C.J. Sawio. Ottawa: IDRC.

Maxwell, D., C. Levin, M. Armar-Klemesu, M. Ruel, and S. Morris. 2000. *Urban livelihoods and food and nutrition security in Greater Accra, Ghana.* Washington, DC, US: IFPRI.

Maye, D., and J. Kirwan. 2010. Alternative food networks. *Sociology of Agriculture and Food* 20: 383–389.

Maye, D., L. Holloway, and M. Kneafsey (eds.). 2007. *Alternative food geographies: Representation and practice.* Oxford: Elsevier.

Mayer, P. (1971). *Townsmen Or tribesmen: Conservatism and the process of urbanization in a South African city.* Oxford University Press.

Mbaye, A., and P. Moustier. 2000. Market-oriented urban agricultural production in Dakar. *Growing cities, growing food: Urban agriculture on the policy agenda*, 235–256.

Mbiba, B. 1995. *Urban agriculture in Zimbabwe: Implications for urban management and poverty.* Aldershot: Avebury.

Mbiba, B. 2000. Urban agriculture in Harare: Between suspicion and repression. In *Growing cities, growing food: Urban agriculture on the policy agenda—A reader on urban agriculture*, ed. N. Bakker, M. Dubbeling, S. Gündel, U. Sabel-Koschella, and H. de Zeeuw, 285–301. Feldafing: GTZ.

Mbiba, B. 2001. The marginalisation of urban agriculture in Lusaka. *Urban Agriculture Magazine* 4: 19–21.

McCarthy, J. 1997. Revitalization of the core city: The case of Detroit. *Cities* 14 (1): 1–11.

McClintock, N. 2014. Radical, reformist, and garden-variety neoliberal: Coming to terms with urban agriculture's contradictions. *Local Environment: The International Journal of Justice and Sustainability* 19 (2): 147–171.

McKernan, M. 1983. All In! Australia during the second world war. *Nelson.*

Medearis, D., and W. Daseking. 2012. Freiburg, Germany: Germany's eco-capital. In *Green cities of Europe*, 65–82. Washington, DC: Island Press.

Melbourne Planning Authority. 2008. Creating liveable new communities—Promising practice: A book of 'good practice' case studies. http://www.mpa.vic.gov.au/wp-content/Assets/Files/Promising%20Practice%209%20April%2008.pdf. Accessed 1 Feb 2016.

Milan Urban Food Policy Pact. 2015. http://www.milanurbanfoodpolicypact.org.

Mkwela, H.S. 2014. Governance of agriculture in the cities of developing countries: Local leaders' perspectives. *Journal of Geography and Regional Planning* 7 (3): 36.

Mlozi, M.R. 1996. Urban agriculture in Dar es Salaam: Its contribution to solving the economic crisis and the damage it does to the environment. *Development Southern Africa* 13 (1): 47–65.

Miller, C. 2003. In the sweat of our brow: Citizenship in American domestic practice during WWII-victory gardens. *The Journal of American Culture* 26 (3): 395–409.

Mitchell, D. 2003. *The right to the city: Social justice and the fight for public space.* Guilford Press.

Mitlin, D., and J. Mogaladi. 2013. Social movements and the struggle for shelter: A case study of eThekwini (Durban). *Progress in Planning* 84: 1–39.

Mittal, A. 2009. *The 2008 food price crisis: Rethinking food security policies, G-24 DP 56.* Geneva: UNCTAD.

Mougeot, L.J.A. 1994. *Urban food production: Evolution, official support and significance.* Cities Feeding People Report 8. Ottawa: IDRC.

Mougeot, L. 1999. For self-reliant cities: Urban food production in a globalizing south. In *For hunger-proof cities: Sustainable urban food systems,* ed. M. Koc et al. Ottawa: IDRC.

Mougeot, L.J.A. 2000. Urban agriculture: Definition, presence, potentials and risks. In *Growing cities, growing food: Urban agriculture on the policy agenda,* ed. N. Bakker, M. Dubbeling, S. Gündel, U. Sabel-Koschella, H. de Zeeuw. Zentralstelle fuer Ernaehrung und Landwirtschaft, 1–42. Feldafing: German Foundation for International Development.

Mtani, A. 1997. Urban agriculture in Dar es Salaam. *Geographical Journal of Zimbabwe* 28: 49–58.

Mun Bbun, T., and A. Thornton. 2013. A level playing field? Improving market availability and access for small-scale producers in Johannesburg, South Africa. *Applied Geography* 36 (1): 40–48.

Nally, D. 2011. The biopolitics of food provisioning. *Transactions of the Institute of British Geographers* 36 (1): 37–53.

National Farmer Federation. 2017. http://www.nff.org.au/farm-facts.html.

Ncube, N., and D. Ncube. 2016. Urban agriculture and food security: A case study of Old Pumula suburb of Bulawayo in Zimbabwe. *Global Journal of Advanced Research* 3 (8): 771–782.

Nel, E., G. Hampwaye, A. Thornton, C. Rogerson, and L. Marais. 2009. Institutional responses to decentralization, urban poverty, food shortages and urban agriculture. GDN Working Paper Series, Working Paper No. 36, Global Development Network (GDN), 1–29.

Nicholls, W.J. 2008. The urban question revisited: The importance of cities for social movements. *International Journal of Urban and Regional Research* 32 (4): 841–859.

Novo, M.G., and C. Murphy. 2000. Urban agriculture in the city of Havana: A popular response to a crisis. In *Growing cities, growing food. Urban agriculture on the policy agenda,* ed. N. Bakker, M. Dubbeling, S. Gündel, U. Sabel-Koshella, and H. de Zeeuw, 329–346. Feldafing, Germany: Zentralstelle für Ernährung und Landwirtschaft (ZEL).

Nugent, R. 2000. The impact of urban agriculture on the household and local economies. In *Growing cities, growing food: Urban agriculture on the policy agenda*, ed. N. Bakker et al. Feldafing: Deutsche Stiftung fuer internationale Entwicklung.

Nunan, F. (2000). Waste recycling through urban farming in Hubli-Dharwad. In *Growing cities, growing food: Urban agriculture on the policy agenda*, ed. N. Bakker, M. Dubbeling, S. Guendel, U. Sabel-Koaschella, and H. Zeeuw de. Eurasburg: DSE.

O'Hara, S.U., and S. Stagl. 2001. Global food markets and their local alternatives: A socio-ecological economic perspective. *Population and Environment* 22 (6): 533.

Okereke, C. 2007. *Global justice and neoliberal environmental governance: Ethics, sustainable development and international co-operation.* London: Routledge.

Oosterveer, P., and D.A. Sonnenfeld. 2012. *Food, globalization and sustainability.* London: Routledge.

Opitz, I., R. Berges, A. Piorr, and T. Krikser. 2016. Contributing to food security in urban areas: Differences between urban agriculture and peri-urban agriculture in the Global North. *Agriculture and Human Values* 33 (2): 341–358.

Orsini, F., R. Kahane, R. Nono-Womdim, and G. Gianquinto. 2013. Urban Agriculture in the developing world: A review. *Agronomy for Sustainable Development* 33 (4): 695–720.

Pack, C.L. 1917. Urban and suburban food production. *Annals of the American Academy of Political and Social Science* 74: 203–206.

Pack, C.L. 1919. *The war garden victorious.* Philadelphia: J.B. Lippincott Company.

Parnell, S., and J. Robinson. 2012. (Re)theorizing cities from the global south: Looking beyond neoliberalism. *Urban Geography* 33 (4): 593–617.

Pierre, J. 2011. *The politics of urban governance.* New York: Palgrave Macmillan.

Pires, V. 2011. Planning for urban agriculture planning in Australian cities. State of Australian Cities.

Popovitch, T. 2014. 10 American cities lead the way with urban agriculture ordinances. Available at http://seedstock.com/2014/05/27/10-american-cities-lead-the-way-with-urban-agriculture-ordinances/. Accessed 29 May 2014.

Potter, R., and T. Unwin (eds.). 1989. *The geography of urban-rural interaction in developing countries.* London: Routledge.

Potutan, G.E., W.H. Schnitzler, J.M. Arnado, L.G. Janubas, and R.J. Holmer. (2000). Urban agriculture in Cagayan de Oro: A favourable response of city government and NGOs. *Growing cities, growing food: urban agriculture on the policy agenda*, 413–428.

Poulsen, M.N., P.R. McNab, M.L. Clayton, and R.A. Neff. 2015. A systematic review of urban agriculture and food security impacts in low-income countries. *Food Policy* 55: 131–146.

Prain, G., N. Gonzales, B. Arce, and J. Tenorio. 2010. Organic vegetable production on the peri-urban interface: Helping low income producers access high value markets in Lima, Peru. *Acta Horticulturae (ISHS)* 881: 117–123.

Purcell, M. 2002. Excavating Lefebvre: The right to the city and its urban politics of the inhabitant. *GeoJournal* 58 (2–3): 99–108.

Purnomohadi, N. 2000. Jakarta: Urban agriculture as an alternative strategy to face the economic crisis. In *Growing cities, growing food. Urban agriculture on the policy agenda,* ed. N. Bakker, M. Dubbeling, S. Gündel, U. Sabel-Koshella, and H. de Zeeuw, 453–465. Feldafing, Germany: Zentralstelle für Ernährung und Landwirtschaft (ZEL).

Quick, S. 1977. Bureaucracy and rural socialism in Zambia. *The Journal of Modern African Studies* 15 (3): 379–400.

Rakodi, C. 1988. Urban agriculture: Research questions and Zambian evidence. *The Journal of Modern African Studies* 26 (3): 495–515.

Resnick, D. 2014. Urban governance and service delivery in African cities: The role of politics and policies. *Development Policy Review* 32 (s1).

Richards, R., and S. Taylor. 2012. Changing land use on the periphery: A case study of urban agriculture and food gardening in Orange Farm. The South African Research Chair in Spatial Analysis and City Planning. Johannesburg: The University of the Witwatersrand. http://hdl.handle.net/10539/17142.

Roberts, B. 2007. Changes in urban density: Its implications on the sustainable development of Australian Cities. In *Proceedings of the State of Australian Cities national conference (SOAC),* 720–739, November 28–30, Adelaide, Australia.

Rogerson, C. 1992. Feeding Africa's cities: The role and potential for urban agriculture. *Africa Insight* 22: 229–234.

Rogerson, C.M. 1993. Urban agriculture in South Africa: Policy issues from the international experience. *Development Southern Africa* 10 (1): 33–44.

Rogerson, C. 2003. Towards 'pro-poor' urban development in South Africa: The case of urban agriculture. *Acta Academica Supplementum* 1: 130–158.

Rosol, M. 2010. Public participation in post-Fordist urban green space governance: The case of community gardens in Berlin. *International Journal of Urban and Regional Research* 34 (3): 548–563.

Roseland, M. 2012. *Toward sustainable communities: Solutions for citizens and their governments,* 4th ed. Gabriola Island, BC: New Society Publishers.

Rosin, C., P. Stock, and H. Campbell (eds.). 2012. *Food systems failure: The global food crisis and the future of agriculture.* London: Earthscan.

Rosin, C., P. Stock, and H. Campbell (eds.). 2013. *Food systems failure: The global food crisis and the future of agriculture.* London: Routledge.

Rosset, P., and D. Benjamin. 1994. *The greening of Cuba: A national experiment in organic agriculture.*

Sabiiti, E.N., and C.B. Katongole. 2014. Urban agriculture: A response to the food supply crisis in Kampala City, Uganda. In *The security of water, food, energy and liveability of cities*. Water Science and Technology Library, ed. B. Maheshwari, R. Purohit, H. Malano, V. Singh, and P. Amerasinghe, vol. 71. Dordrecht: Springer.

Sahn, D.E. 1989. A conceptual framework for examining the seasonal aspects of household food security. In *Seasonal variability in third world agriculture: The consequences for food security*, ed. D.E. Sahn. Baltimore: John Hopkins University Press.

Sanyal, B. 1985. Urban agriculture: Who cultivates and why? A case-study of Lusaka, Zambia. *Food and Nutrition Bulletin* 7: 15–24.

Sanyal, B. 1987: Urban cultivation amidst modernisation: How should we interpret it? *Journal of Planning Education and Research* 6: 187–207.

Säumel, I., I. Kotsyuk, M. Hölscher, C. Lenkereit, F. Weber, and I. Kowarik. 2012. How healthy is urban horticulture in high traffic areas? Trace metal concentrations in vegetable crops from plantings within inner city neighbourhoods in Berlin, Germany. *Environmental Pollution* 165: 124–132.

Sawio, C. 1994. Who are the farmers of Dar es Salaam? In *Cities feeding people: An examination of urban agriculture in East Africa*, ed. A.G. Egziabher, et al., 23–44, Ottawa: International Development Research Centre.

Schlosser, E. 2002. *Fast food nation: The dark side of the All-American meal*. Houghton Mifflin.

Sidaway, J.D. 2000. Postcolonial geographies: An exploratory essay. *Progress in Human Geography* 24 (4): 591–612.

Simon, D. 1995. Debt, democracy and development: Sub-Saharan Africa in the 1990s. In *Structurally adjusted Africa: Poverty, debt and basic needs*, ed. D. Simon, W. van Spengen, C. Dixon, and A. Närman. Pluto, London.

Smit, J., A. Ratta, and J. Nasr. 1996. *Urban agriculture: Food, jobs and sustainable cities*. New York: UNDP.

Smit, J., A. Ratta, and J. Nasr. 2001. *Urban agriculture: Food, jobs and sustainable cities*, 2001st ed. New York: UNDP.

Smith, D., and D. Tevera. 1997. Socio-economic context for the householder of urban agriculture in Harare, Zimbabwe. *Geographical Journal of Zimbabwe* 28: 25–38.

Smith, A., and J.B. McKinnon. 2007. *The 100-mile Diet: A year of local eating*. Toronto: Random House Canada.

Sneyd, L., A. Legwegoh, and E. Fraser. 2013. Food riots: Media perspectives on the causes of food protest in Africa. *Food Security* 5 (4): 485–497.

Soja, E. 2010. Spatialising the urban. Part 1. *City: Analysis of Urban Trends, Culture, Theory, Policy, Action* 14 (6): 629–635.

South African Government. 2012. National Development Plan—2030, Ch 8–9. Available at: http://www.gov.za/sites/www.gov.za/files/devplan_ch8_0.pdf. Accessed 28 Sept 2017.

Tacoli, C. 1998. Rural-urban interactions: A guide to the literature. *Environment and Urbanization* 10 (1): 147–166.

Tawodzera, G. 2011. Vulnerability in crisis: Urban household food insecurity in Epworth, Harare, Zimbabwe. *Food Security* 3 (4): 503–520.

Thaman, Randy. 1975. *Urban gardening in Papua, New Guinea and Fiji: Present status and implications for urban land-use planning.* Suva: University of the South Pacific.

Thornton, A. 2008. Beyond the metropolis: Small town case studies of urban and peri-urban agriculture in South Africa. *Urban Forum* 19 (3): 243–262.

Thornton, A. 2011. Food for thought? The potential of urban agriculture in local food production for food security in the South Pacific. In *Food systems failure: The global food crisis and the future of agriculture*, ed. C. Rosin, P. Stock, and H. Campbell, 203–218. London: Earthscan.

Thornton, A. 2012. *Urban agriculture in South Africa: A study of the Eastern Cape Province.* Lewiston, NY: Edwin Mellen Press.

Thornton, A. 2017. "The Lucky country"? A critical exploration of community gardens and city–community relations in Australian cities. *Local Environment* 22 (8): 969–985.

Thornton, A. 2018. Food security in African Cities. In *The Routledge Handbook of African development*, ed. T. Binns, K. Lynch, and E. Nel. London: Routledge.

Thornton, A., and C. Rogerson. 2013. African cities and the millennium development goals: A case for applied geography. *Applied Geography* 36 (1): 1–3.

Thornton, A., E. Nel, and G. Hampwaye. 2010. Cultivating Kaunda's plan for self-sufficiency: Is urban agriculture finally beginning to receive support in Zambia? *Development Southern Africa* 27 (4): 613–625.

Thornton, A., J. Momoh, and P. Tengbe. 2012. Institutional capacity building for urban agriculture research using Participatory GIS in a post-conflict context: A case study of Sierra Leone. *Australasian Review of African Studies* 33 (1): 165–176.

Thornton, A., K. Lyons, and S. Sharpe. 2018. Community gardens for social and food justice: The case of urban agriculture in Australian cities. In *The Routledge handbook of community development research*, ed. L. Shevellar and P. Westoby. London: Routledge.

Tidball, K.G., and M. Krasny (eds.). 2014. *Greening in the red zone: Disaster, resilience and community greening.* Dordrecht: Springer.

Tilly, C. 1999. From interactions to outcomes in social movements. In *How social movements matter*, ed. M. Giugni, D. McAdams, and C. Tilly. Minneapolis: University of Minnesota Press.

Tisdell, C.A. 2013. *Competition, diversity and economic performance.* Cheltenham: Edward Elgar Publishing.

Torres-Lima, P., L.M. Rodriques Sanchez, and B.I. Garcia Uriza. 2000. Mexico City: The integration of urban agriculture to contain urban sprawl. In

Growing cities, growing food. Urban agriculture on the policy agenda, ed. N. Bakker, 363–390. Germany: DSE.

UN-Habitat. 2015. Governance. Available at: http://unhabitat.org/urban-themes/governance/. Accessed 20 Oct 2017.

United Nations, Department of Economic and Social Affairs (DESA), Population Division. 2015. World urbanisation prospects: The 2014 Revision. Report No. ST/ESA/SER.A/366.

United Nations Development Programme (UNDP). (2015a). The post-2015 development agenda. Available at http://www.undp.org/content/undp/en/home/mdgoverview/post-2015-development-agenda.html. Accessed 5 Dec 2017.

UNDP. (2015b). SDG Goal 11: Cities will play an important role in achieving the SDGs. Available at http://www.my.undp.org/content/malaysia/en/home/presscenter/articles/2015/08/04/sdg-goal-11-cities-will-play-an-important-role-in-achieving-the-sdgs.html. Accessed 5 Dec 2017.

Van Veenhuizen, R., G. Prain, and H. De Zeeuw. 2001. Appropriate methods for urban agriculture research, planning implementation and evaluation. *Urban Agriculture Magazine* 5: 1–5.

Venn, L., M. Kneafsey, L. Holloway, R. Cox, E. Dowler, and H. Tuomainen. 2006. Researching European 'alternative' food networks: Some methodological considerations. *Area* 38 (3): 248–258.

Von Bertalanffy, L., and A. Rapoport (eds.). 1963. *General systems*. Ann Arbor: Society for General Systems Research.

Von Thünen, J.H. 1875. In *Der isolirte staat in beziehung auf landwirtschaft und nationalökonomie*, vol. 1. Wiegant, Hempel & Parey.

Walker, R.E., C.R. Keane, and J.G. Burke. 2010. Disparities and access to healthy food in the United States: A review of food deserts literature. *Health & Place* 16 (5): 876–884.

Warner, S. 1987. *To dwell is to garden: A history of Boston's community gardens*. Boston: Northeastern University Press.

Warshawsky, D. 2014. Civil society and urban food insecurity: Analyzing the roles of local food organizations in Johannesburg. *Urban Geography* 35 (1): 109–132.

Warshawsky, D. 2016. Civil society and the governance of urban food systems in sub-Saharan Africa. *Geography Compass* 10 (7): 293–306.

Weiss, E. 2008. *Fruits of victory: The woman's land army of America in the great war*. Washington, DC: Potomac Press.

Wenban-Smith, H., A. Faße, and U. Grote. 2016. Food security in Tanzania: The challenge of rapid urbanisation. *Food Security* 8 (5): 973–984.

White, C., K. Woodfield, J. Ritchie, and R. Ormston. 2014. Writing up qualitative research. In *Qualitative research practice: A guide for social science students and researchers*, 367–400, ed. J. Ritchie, J. Lewis, C.M. Nicholls, and R. Ormston. London: Sage.

Wittman, H., A.A. Desmarais, and N. Wiebe. 2010. The origins and potential of food sovereignty. In *Food sovereignty: Reconnecting food, nature and community*, ed. H. Wittman, A.A. Desmarais, and N. Wiebe, 1–14. Oakland, CA: Food First.

Wolf, K.L. 2010. Crime and fear—A literature review. In *Green cities: Good health*. College of the Environment. Seattle: University of Washington.www.greenhealth.washington.edu

World Resources Institute. 2000. *A guide to world resources 2000–2001, people and ecosystems: The fraying web of life*. Washington DC: World Resources Institute.

Yang, Z., J. Cai, and R. Sliuzas. 2010. Agro-tourism enterprises as a form of multi-functional urban agriculture for peri-urban development in China. *Habitat International* 34 (4): 374–385.

Yi-Zhang, C., and Z. Zhangen. 2000. Shanghai: Trends towards specialised and capital-intensive urban agriculture. *Growing cities growing food: Urban agriculture on the policy agenda*, 467–477. Feldafing: DSE-ZEL.

Yoveva, A., B. Gocheva, G. Voykova, B. Borissov, and A. Spassov. 2000. Sofia: Urban agriculture in an economy in transition. *Growing cities, growing food: Urban agriculture on the policy agenda. A reader on urban agriculture*, 501–518.

Zezza, A., and L. Tasciotti. 2010. Urban agriculture, poverty, and food security: Empirical evidence from a sample of developing countries. *Food Policy* 35 (4): 265–273.

INDEX

© The Editor(s) (if applicable) and The Author(s) 2018
A. Thornton, *Space and Food in the City*,
https://doi.org/10.1007/978-3-319-89324-2